Journal of Applied Logics - IfCoLog Journal of Logics and their Applications

Volume 10, Number 6

December 2023

Disclaimer

Statements of fact and opinion in the articles in Journal of Applied Logics - IfCoLog Journal of Logics and their Applications (JALs-FLAP) are those of the respective authors and contributors and not of the JALs-FLAP. Neither College Publications nor the JALs-FLAP make any representation, express or implied, in respect of the accuracy of the material in this journal and cannot accept any legal responsibility or liability for any errors or omissions that may be made. The reader should make his/her own evaluation as to the appropriateness or otherwise of any experimental technique described.

ISBN 978-1-84890-444-6
ISSN (E) 2631-9829
ISSN (P) 2631-9810

College Publications
Scientific Director: Dov Gabbay
Managing Director: Jane Spurr

`http://www.collegepublications.co.uk`

Editorial Board

Scope and Submissions

This journal considers submission in all areas of pure and applied logic, including:

pure logical systems
proof theory
constructive logic
categorical logic
modal and temporal logic
model theory
recursion theory
type theory
nominal theory
nonclassical logics
nonmonotonic logic
numerical and uncertainty reasoning
logic and AI
foundations of logic programming
belief change/revision
systems of knowledge and belief
logics and semantics of programming
specification and verification
agent theory
databases

dynamic logic
quantum logic
algebraic logic
logic and cognition
probabilistic logic
logic and networks
neuro-logical systems
complexity
argumentation theory
logic and computation
logic and language
logic engineering
knowledge-based systems
automated reasoning
knowledge representation
logic in hardware and VLSI
natural language
concurrent computation
planning

This journal will also consider papers on the application of logic in other subject areas: philosophy, cognitive science, physics etc. provided they have some formal content.

Submissions should be sent to Jane Spurr (jane@janespurr.net) as a pdf file, preferably compiled in LaTeX using the IFCoLog class file.

Contents

ARTICLES

Editorial Note for the Special Issue on Multiple Valued Logic 2023

Martin Lukac
Hiroshima City University, Japan
martin.lukac@nu.edu.kz

In this issue special issue on Multiple-Valued Logic with original contributions as well as with extended contributions from the International Symposium on Multiple-Valued Logic (ISMVL) 2022 we are delighted to present several works of high quality and exciting content. As we are slowly closing towards an imminent change of technology in the computing, the drastic change ahead will also radically speed up a change in various related areas. Logic, mathematics, complexity and other areas with marginal fields of interests will become of prominent inters with the emergence of quantum computing.

As is usual we have papers in three different areas: circuits, computing and algorithms and theory. All papers propose advanced knowledge in their respective areas of research in the Multuiple-Valued Logic and are well aligned with the expected arrival of the quantum computing platforms and commercial applications.

In the area of circuits, devices and signal, the first paper entitled *Multi-Valued Data Transmission Quality Evaluation Using Two-Dimensional PAM-4 Symbol Mapping* by Kazuharu Nakajima, Yasushi Yuminaka and Yosuke Iijima from Gunma University and Oyama College, Japan. The paper presents work in the 2D symbol transition mapping during the eye-opening signal monitoring in a pulse amplitude modulation with four levels. The paper shows that the eye-opening monitor is sufficient to assess the quality of the transmission in the PAM-4 regime by allowing to determine the intersymbol interference.

In the area of computing and algorithms, the paper entitled *Quantum Algorithms for Unate and Binate Covering Problems with Application to Finite State Machine Minimization* by Abdirahman Alasow and Marek Perkowski from Portland State University, USA describes heuristics for designing algorithms on a quantum computer for the covering problems. Covering problem is a constraint satisfaction based logic problems often used in the design of classical logic circuits. The authors demonstrate that it is possible to build quantum algorithms that can accelerate the search for minimal covering and therefore minimize the resulting effort of the designer as when compared to classical methods.

Finally in the theoretical area we have this year four papers. Two of the papers focus purely on logic and mathematical logic while the other two are focusing more on function analysis of specific types of logic functions.

The first paper entitled *Embedding first-order classical logic into Gurevich's extended first-order intuitionistic logic: The role of strong negation* by Norihiro Kamide from Nagoya City University, Japan presents a method of embedding first-order logic to the higher order intuitionistic logic. The paper presents several steps of the described approach by introducing an extension called ELK to the sequent calculus LK and LJ. Then tools for embedding the first-order logic in the ELK framework is embedded into the LJ intuitionistic framework is proposed. The methodology in the paper support the strong negation and thus is well suited as the embedding approach of the first order logic.

The next paper entitled *On Weak Bases for Boolean Relational Clones and Reductions for Computational Problems* by Mike Behrisch from TU Wien, Austria discusses the weak bases on finite sets in the context of relational clones. The paper considers groups of similar clones as a parallel representation to the clones given by the Post's lattice. The main result of this paper is the analysis of weak bases that can be used to reduce computational problems such as satisfability by showing that the instances grow linearly in size.

The third paper in this group is entitled *On Representation of Maximally Asymmetric Functions Based on Decision Diagrams* by Shinobu Nagayama, Tsutomu Sasao and Jon Butler from Hiroshima City University, Japan, Meiji University, Japan and Naval Postgraduate School, USA respectively. The paper discusses the multiple-valued maximally asymmetric functions, functions that are as far as possible from the symmetric functions. Symmetric functions are a class of functions that do not change the output under input variable permutations and thus are useful tools in the design of logic circuits. The paper discusses the usage of maximally asymmetric functions in cryptography and for this purpose the paper studies efficient representation of such functions in order to promote the building of benchmarks for future use.

The last paper in this section and this special issue is entitled *p-valued Maiorana-McFarland Functions Structure of Their Reed-Muller Spectra* by Claudio Moraga, Radomir Stanković and Milena Stanković from TU Dortmund, Germany, Mathematical Institute of SASA, Serbia and University of Niš, Serbia respectively. The paper discusses the existence of specific type of Bent functions for input variable radix $p > 2$. Bent functions are the least linear functions (or the most non linear) and therefore are considered as a good candidate for the usage in cryptography. The paper discusses their representation and identification using the Reed-Muller spectrum unlike the classical representation that uses the flatness of the Walsh spectrum.

We believe that this year's special issue on MVL will be an exciting addition to the study of the latest MVL trends and we hope to see you at the next ISVML event.

Received November 2023

Multi-Valued Data Transmission Quality Evaluation Using Two-Dimensional PAM-4 Symbol Mapping

Kazuharu Nakajima
Graduate School of Science and Technology, Gunma University, JAPAN
T211D053@gunma-u.ac.jp

Yasushi Yuminaka
Graduate School of Science and Technology, Gunma University, JAPAN
yuminaka@gunma-u.ac.jp

Yosuke Iijima
National Institute of Technology (KOSEN), Oyama College, JAPAN
yiijima@oyama.kosen-ac.jp

Abstract

This paper presents an eye-opening monitor technique leveraging four-level pulse amplitude modulation (PAM-4) symbol transition characteristics. This approach facilitates the evaluation of data transmission quality concerning the adaptive coefficient settings of PAM-4 equalizers. The two-dimensional (2D) symbol transition mapping visually depicts the degree of intersymbol interference (ISI). This work expands upon the 2D mapping model introduced in [14], adding more theoretical analysis and new simulation results. Both simulation and empirical results indicate that 2D symbol mapping can assess the quality of PAM-4 data transmission impaired by ISI and can visually represent the equalization effect.

This paper is an extension of [14].
This work was supported by JSPS KAKENHI Grant Numbers 21K11819 and 21H01381.

1 Introduction

The demand for fast data transmission is growing [1], driven by the increasing need for high-speed, large-capacity information communication services like the Internet of Things (IoT), 5G applications, and cloud computing. This surge in demand has significantly increased data transmission volumes at data centers and other facilities. However, when transmitting data at high speeds, electric wires operate as low-pass filters, leading to the degradation of the high-frequency components of the transmitted data. The restricted bandwidth of the channel can cause intersymbol interference (ISI) at the receiving end. Therefore, the utilization of waveform shaping techniques becomes essential to minimize such interference. Multi-valued signaling effectively lowers the Nyquist frequency of the transmitted data, thereby diminishing the effect of ISI. The 4-level pulse amplitude modulation (PAM-4) technique can halve the Nyquist frequency while maintaining the same bitrate as binary data signaling. PAM-4 signaling can transmit double the information compared to binary signaling within the same Nyquist frequency, given that PAM-4 can transmit 2 bits of data in a single symbol.

However, PAM-4 signaling is three times more susceptible to noise amplitude than binary signaling. This sensitivity is due to decreased distances between each symbol, complicating threshold discrimination and impeding accurate determination of PAM-4 symbols. Thus, waveform shaping techniques, such as the continuous-time linear equalizer (CTLE) and/or decision feedback equalizer (DFE), are employed in high-speed data transmission systems to counteract ISI at the receiver's end [2, 3]. A transmitter-side equalizer, like a feed-forward equalizer (FFE), can effectively mitigate the effects of ISI [4]-[6]. These waveform shaping circuits demand parameter adjustments [7], which are based on the transmission characteristics. Determining the actual transmission line characteristics can be complex and difficult due to the influence of adjacent wiring and peripheral components like connectors. Additionally, the transmission environment can fluctuate over time. To handle these complexities, eye-opening monitor (EOM) techniques [8]-[13] are used to assess the quality of received signals under adaptively adjusted equalizer parameters. However, the three eyes of the PAM-4 hinder the implementation of the corresponding EOM algorithm.

Consequently, this paper presents new EOM techniques based on two-dimensional (2D) symbol mapping to visualize the symbol distribution using an analog-to-digital converter (ADC)-based receiver [14]. The 2D symbol mapping enables us to visualize the ISI effect of the transmitted multi-valued symbols [15, 16]. This research investigates 2D symbol mapping to shed light on the relationship between ISI and 2D symbol transition pattern mapping in multi-valued data transmis-

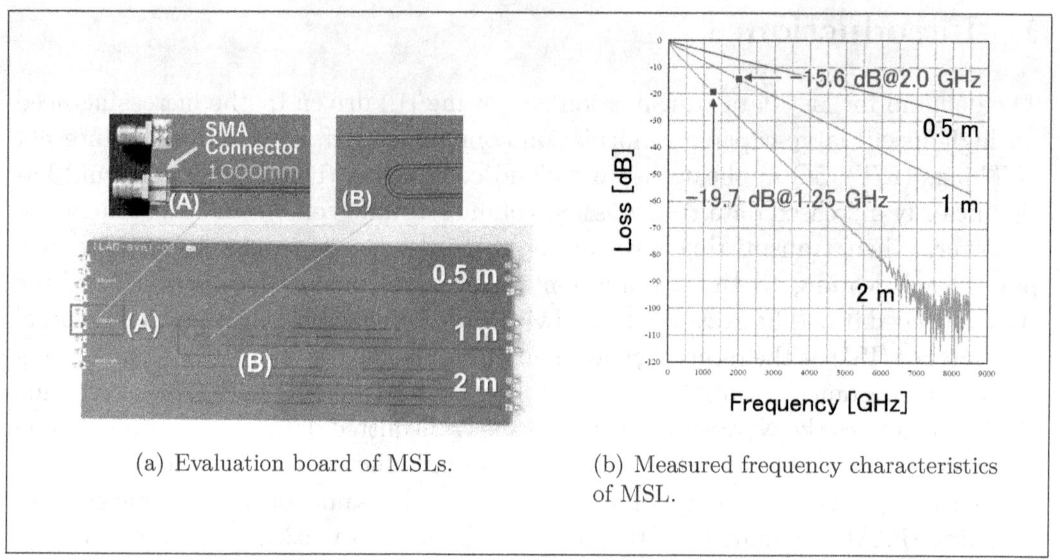

(a) Evaluation board of MSLs.

(b) Measured frequency characteristics of MSL.

Figure 1: Measured frequency characteristics of MSLs (0.5, 1, and 2 m).

sion, taking into account eye-opening conditions. Both simulation and experimental results substantiate the efficacy of 2D symbol mapping in assessing the integrity of PAM-4 data transmission compromised by ISI while also effectively illustrating the impact of equalization. In this paper, we advance and extend the 2D mapping model presented in [14] by integrating a theoretical analysis and additional simulation results. Specifically, we examine how post-cursors and pre-cursors of the impulse response in transmission channels influence the form of 2D mapping.

The structure of this paper is as follows: Section 2 discusses the symbol distribution in multi-valued signaling affected by ISI. The proposed 2D PAM-4 symbol transition mapping method is detailed in Section 3. Section 4 provides simulation and experimental results to validate the proposed method using 2D symbol mapping. The findings are summarized and discussed in Section 5, and finally, Section 6 concludes the paper.

2 Symbol distribution with ISI in PAM-4 signaling

2.1 Symbol distribution of received PAM-4 symbols

In high-speed data transmission over a band-limited channel, the frequency characteristics of the transmission line have an impact on the received symbols. Figure 1 displays the measured frequency characteristics of micro-strip lines (MSLs) that

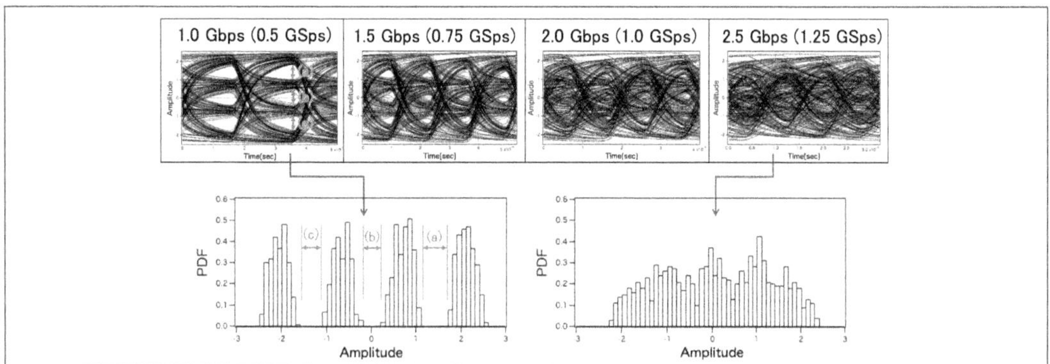

Figure 2: Simulation results of eye diagram of each data rate and symbol distribution of both 1 Gbps and 2.5 Gbps PAM-4 on the 2-m MSL.

were specifically fabricated for evaluation purposes. The lengths of these MSLs are 0.5, 1, and 2 m. The evaluation board is composed of a glass epoxy multilayer PCB, with a dielectric thickness of 0.1 mm and a relative permittivity of 4.1 to 4.2 at 1 GHz, and sub-miniature type A (SMA) connectors. The wiring pattern has a width of 0.2 mm and an 18 μ m thickness. Figure 1(b) shows the S_{21} characteristic measured using a vector network analyzer. The attenuation at 2 GHz is -15.6 dB in the 1-m MSL, and the attenuation at 1.25 GHz is -19.7 dB in the 2-m MSL.

The eye diagrams of the PAM-4 data transmission on the 2-m MSL are depicted in Fig. 2. In these diagrams, we can observe the impact of ISI on both vertical eye-opening and horizontal eye-opening. The ISI effect manifests as a decrease in the amplitude of the vertical eye-opening and as timing ambiguity in the horizontal eye-opening. The eye is open at a PAM-4 data rate of 1 Gbps, and each symbol distribution is well-separated, as illustrated in Figs. 2(a)–(c). However, the eye becomes completely closed due to severe ISI when the data rate increases to 2.5 Gbps. In this scenario, the Nyquist frequency is 1.25 GHz, considering a channel loss of -19.7 dB. Consequently, the detection of each symbol becomes challenging because the ISI effect causes overlapping symbol distributions, leading to a loss of distinction between symbols.

2.2 Estimation of symbol distribution using Gaussian mixture model

Various eye-opening monitor techniques have been proposed to evaluate the ISI effect at the receiver [8]-[13]. Our previous proposal involved an eye-opening monitor technique that utilized a Gaussian mixture model (GMM) estimation approach

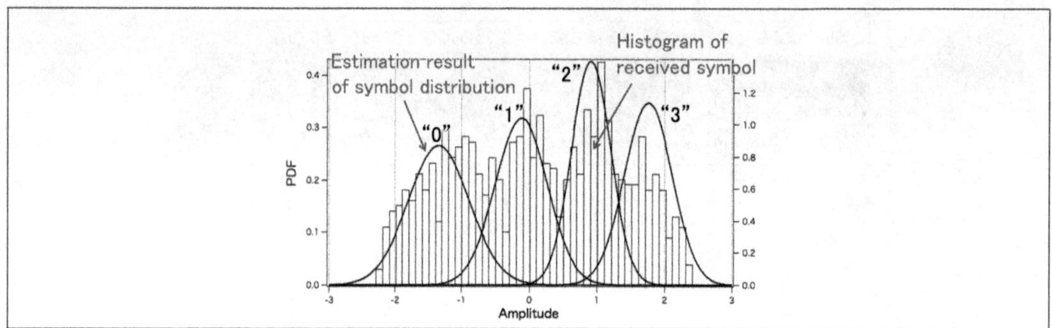

Figure 3: GMM estimation simulation results of symbol distribution of 2.5 Gbps PAM-4 on the 2-m MSL.

to evaluate symbols while the eye is closed [9]-[13]. Within the GMM estimation method, the distribution of each symbol can be determined by fitting a curve to the GMM, assuming that the received symbols follow a Gaussian distribution. In the case of PAM-4, the expression for the received symbol distribution can be given as:

$$p(x) = \sum_{k=0}^{3} \pi_k N(x|\mu_k, \sigma_k), \tag{1}$$

where $N(x|\mu_k, \sigma_k)$ represents a Gaussian distribution with a mean μ_k and standard deviation σ_k. Additionally, π_k denotes the mixing coefficient corresponding to the weight of each Gaussian distribution.

The GMM estimation can be achieved by generating four Gaussian function curves, each corresponding to a symbol distribution in PAM-4, as depicted in Fig. 3. The bar graph illustrates the probability density function (PDF) of the received symbols for a 2.5 Gbps PAM-4 signal under a 2-m MSL condition. Through GMM estimation, we can assess the histogram of the received symbol and its associated ISI effect for each symbol, even when the eye is closed.

3 2D mapping for PAM-4 signaling

We have also introduced a symbol-evaluation technique based on 2D PAM-4 symbol transition mapping as an alternative method for evaluating ISI [14]. In this approach, the received PAM-4 symbols ($\ldots$, s_{i-1}, s_i, s_{i+1}, $\ldots$) are plotted on a 2D surface, where the x-axis represents the previous symbol values s_{i-1} and the y-axis represents the current symbol values s_i, as illustrated in Fig. 4. When there is no ISI or attenuation on the transmission line, the transmitted symbols (-3, -1, $+1$, $+3$) are

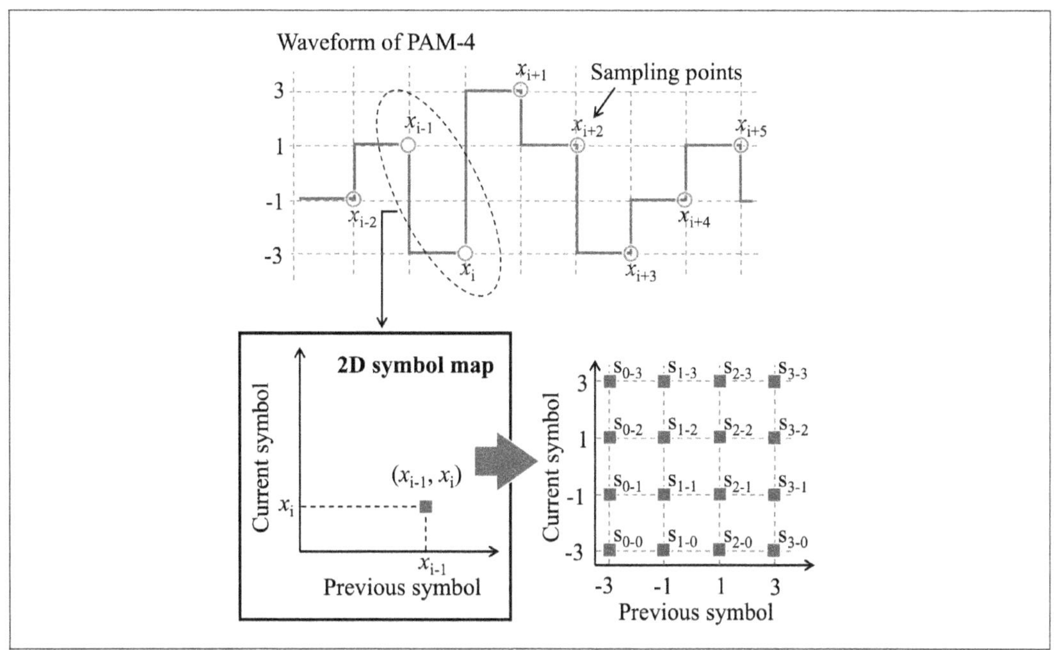

Figure 4: Overview of 2D mapping of PAM-4 received symbols.

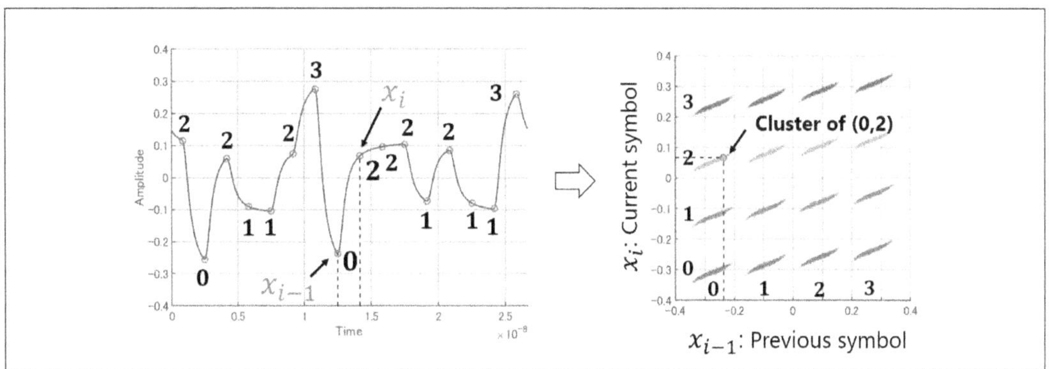

Figure 5: Example of 2D mapping of PAM-4 received symbols.

mapped onto a 2D plane, as depicted in Fig. 4. The transitions between s_{i-1} and s_i are plotted as a 16-point grid pattern.

In contrast, Figure 5 illustrates an example of the 2D mapping of received PAM-4 symbols after transmission through a channel, including the presence of ISI. For instance, if the current symbol value x_i is 2 and the previous symbol x_{i-1} is 0, the corresponding 2D mapping plot would represent the cluster of (0,2), as depicted in

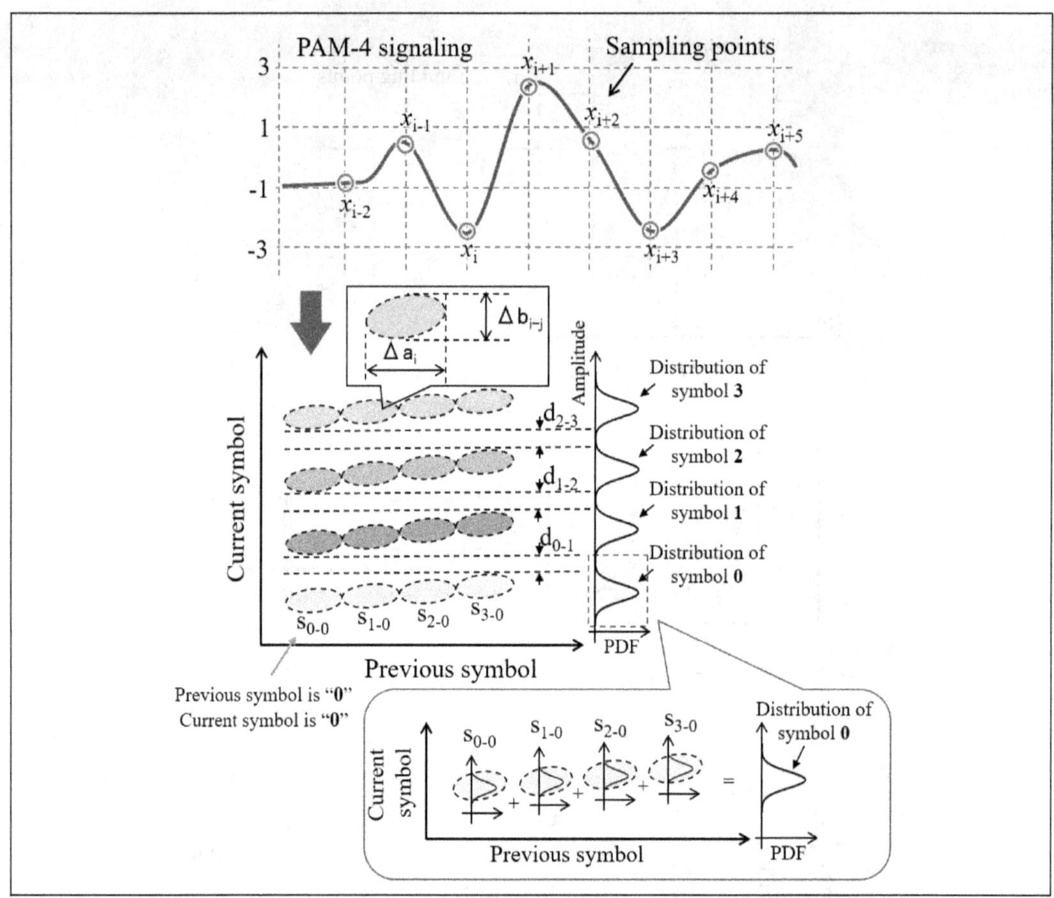

Figure 6: Correlation between 2D symbol mapping with symbol distribution curves.

Fig. 5.

The ISI effect expands and tilts the distributions of the received symbols. As shown in Fig. 6, Δa_i shows the distribution of the previous symbols i, and Δb_{i-j} shows the distribution of symbols of transition symbol i to symbol j. All the symbol distribution curves were derived by projecting these symbol distributions onto the 2D mapping surface, as depicted in Fig. 6. Each symbol distribution curve represents the integration of four distributions resulting from the four transition patterns in PAM-4. For instance, the distribution curve of symbol **3** encompasses the symbol distributions of s_{0-3}, s_{1-3}, s_{2-3}, and s_{3-3}. Within the 2D symbol mapping, the spaces d_{0-1}, d_{1-2}, and d_{2-3} represent the symbol distances. When $d_{i-j} > 0$, the distribution curves of each symbol do not overlap, indicating an open eye. Hence, the utilization of 2D symbol mapping enables the visualization of individual symbol

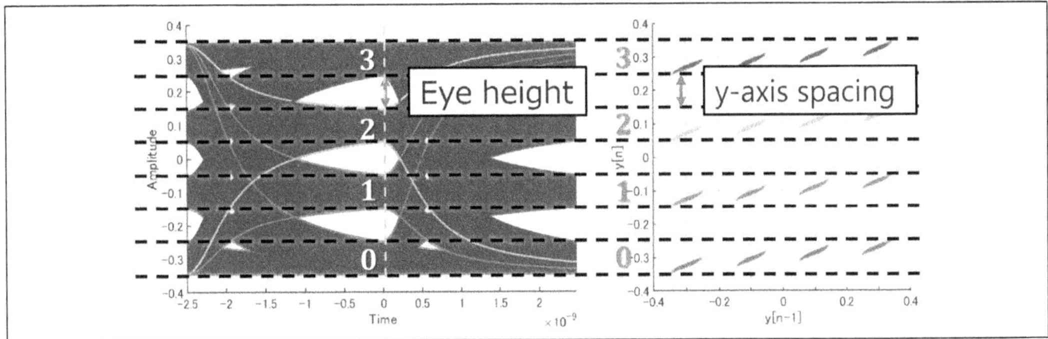

Figure 7: Relationship between eye diagram and 2D mapping.

distributions, which accurately represents the intricate symbol transition patterns involved in multi-valued data transmission. This approach provides an actual measurement rather than relying on predictions from GMM estimation.

Figure 7 illustrates the relationship between the 2D mapping and the conventional eye diagram utilized in waveform degradation evaluation methods. In this context, the spacing along the y-axis in the 2D mapping corresponds to the eye height of the eye aperture at the sampling point of the eye diagram. As the data rate increases and the eye diagram closes, the 2D mapping point cloud exhibits a steeper slope and narrower spacing along the y-axis direction.

4 Simulation and experimental results

The simulation results of the 2D mapping and PDF histogram of the received PAM-4 symbols on a 2-m MSL are depicted in Fig. 8. For this simulation, the impulse response of the transmission line was obtained from measurement results of its transmission characteristics. The received waveform was calculated through numerical simulation using the IGOR software tool, employing the convolution of the data and impulse response. The ADC resolution is neglected in the simulation.

Under the condition of low ISI impact at 1 Gbps PAM-4, each symbol is individually arranged in the 2D mapping, resulting in separated histograms for each symbol, as demonstrated in Fig. 8(a). The histogram of symbol **3** represents the combination of s_{0-3}, s_{1-3}, s_{2-3}, and s_{3-3}. However, as the data rate increases to 2.5 Gbps PAM-4, both Δa_i and Δb_{i-j} expand because of the heightened ISI and the overlapping distributions of current symbols, as illustrated in Fig. 8(b). The tilt of each distribution becomes significantly more pronounced compared to the results at 1 Gbps PAM-4. Consequently, the histogram of the received symbol, which

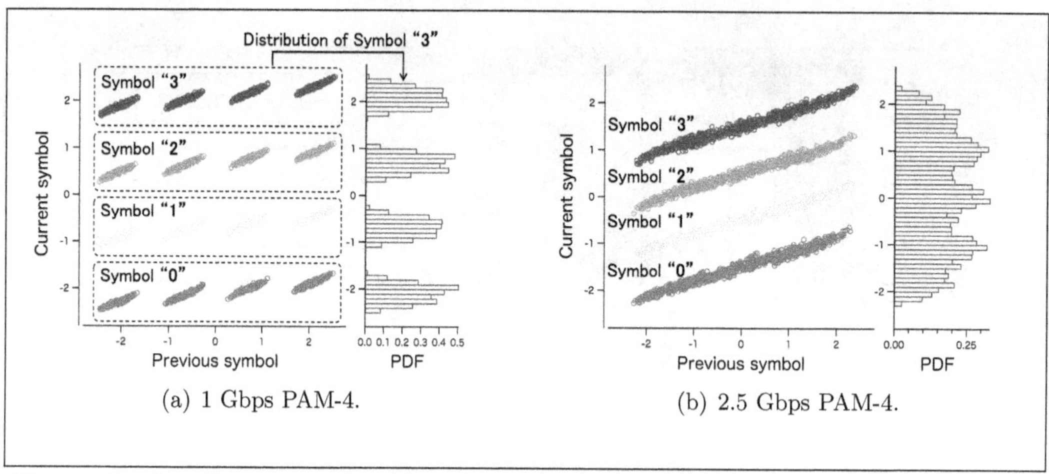

Figure 8: Simulation results of both the 2D mapping and histogram of PAM-4 received symbols with ISI on the 2-m MSL.

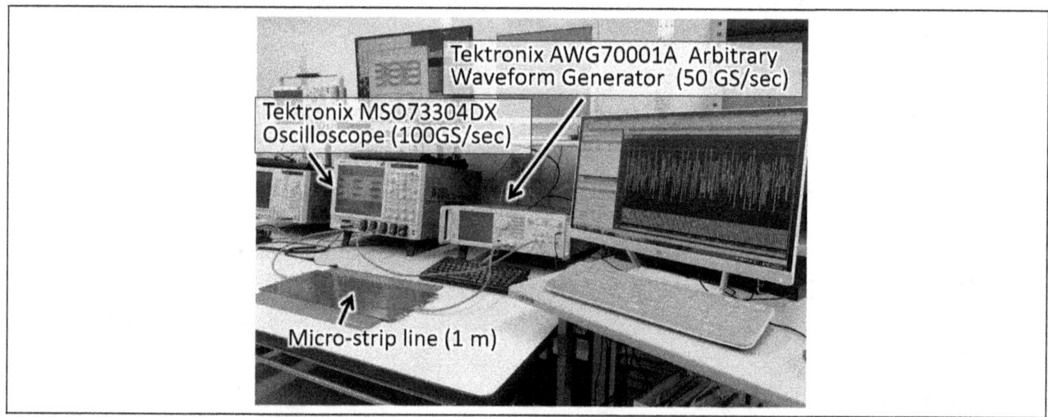

Figure 9: Experimental setup.

represents the projection on the y-axis, combines into a single group. While the 2D symbol mapping visually represents the ISI effect on each symbol transition, symbol classification becomes challenging when using a 1D plot like a histogram, as observed in Fig. 8(b).

The measurement setup for obtaining PAM-4 sampling signals using an arbitrary waveform generator and an oscilloscope is illustrated in Fig. 9. In Fig. 10, the measurement result of a 4 Gbps PAM-4 signal on a 1-m MSL is presented through an eye diagram and a 2D symbol mapping. The heat map distribution is generated from $3,999,994$ sampling points at the receiver. Despite the complete closure of

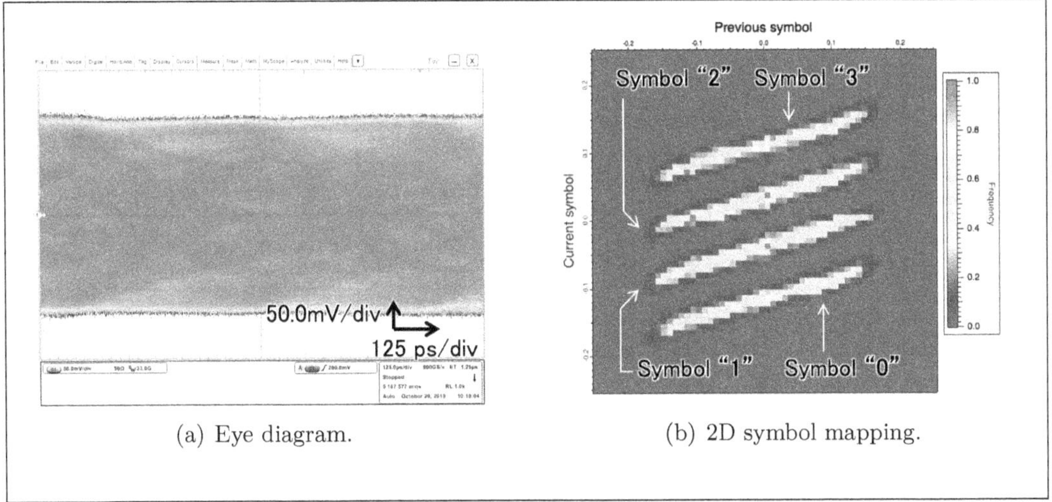

(a) Eye diagram.

(b) 2D symbol mapping.

Figure 10: Evaluation result of 2D symbol mapping using measurement result of received symbol at 4 Gbps PAM-4 on the 1-m MSL.

the eye due to overlapping projections of each symbol on the y-axis, the 2D symbol mapping successfully separates the distribution of each symbol, as depicted in Fig. 10(b). The minimum and maximum ranges are -0.25 and $+0.25$, respectively. With 60×60 pixels in Fig. 10(b), each pixel represents the number of symbols falling within a range of 0.5/60 V. Consequently, this is equivalent to an evaluation using an ADC with a resolution of $0.5/60 = 0.0083$ V. Creating a 1D histogram of the distribution of receiving-end symbols to determine the overlapping symbols proves to be challenging. Conversely, the 2D mapping technique successfully separates each symbol.

5 Discussion

5.1 Evaluation of ISI effect using 2D mapping

The 2D mapping technique preserves the ISI characteristic, even when the eye diagram is closed, enabling the evaluation of the ISI effect, as demonstrated in Fig. 10. Although the distribution of the symbol transition shown in Fig. 10(b) overlaps, the distributions of other symbol transitions do not overlap. As a result, four distinct clusters can be identified. However, capturing the property of this distribution between symbols necessitates sampling at the appropriate time in the received waveform at the receiving end.

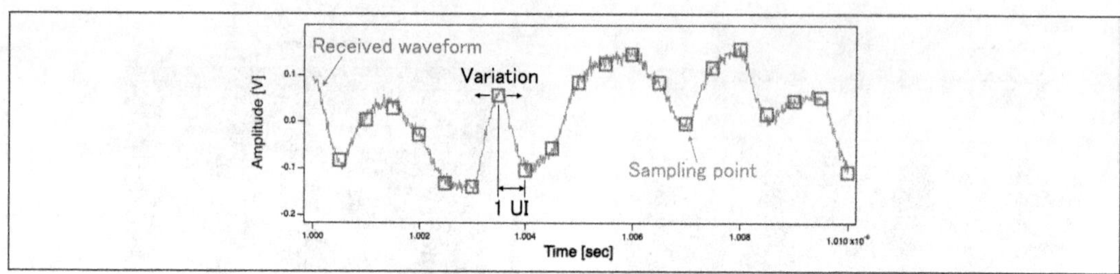

Figure 11: Experimental result of received waveform and sampling points.

In particular, it is crucial to sample each symbol at its peak value in the received waveform. If the sampling timing is incorrect, the characteristic property of the 2D symbol plot is not preserved. Figure 11 illustrates an example case where the sampling timing varies, showing the sampling points in the received waveform of a 4 Gbps PAM-4 signal on a 1-m MSL. Figures 12(a)-(e) display the 2D plot results for the sampling points depicted in Fig. 11, with an error ranging from 5% to 30% of the unit interval (UI) time. As the variation in sampling points increases, the variation of each symbol on the 2D plot also increases, leading to the loss of characteristic properties between symbols due to the ISI.

Under certain conditions, the receiving-end symbol can be identified even when the eye diagram is completely closed due to the distinct characteristics shared between symbols. For example, in Fig. 13, it is demonstrated that, particularly within the overlapping range of specific symbols, the correct symbol can be identified by considering the value of a preceding symbol, even if d_{i-j} is negative. In the example(Fig. 14), the overlapping symbol distribution is determined by whether the previous symbol is to the right or left of the center [13]. However, if multiple overlapping symbol distributions exist, it becomes challenging to determine the correct symbol based on the information from just one previous symbol value.

The slope of the 2D mapping point cloud can be determined by considering the impulse response of the transmission line. In the case where the transmitted symbol value is denoted as $a[n]$ and the received symbol value is denoted as $y[n]$, their relationship can be expressed as follows:

$$\begin{aligned} y[n-1] &= \cdots + h_1 a[n-2] + h_0 a[n-1] + h_{-1} a[n] + \cdots \\ y[n] &= \cdots + h_1 a[n-1] + h_0 a[n] + h_{-1} a[n+1] + \cdots, \end{aligned} \tag{2}$$

where h_i is the sampled value of the impulse response. Approximating Eq. (2) by focusing only on $a[n-1]$ and $a[n]$ yields the following equation:

$$\begin{aligned} y[n-1] &\simeq h_0 a[n-1] + h_{-1} a[n] \\ y[n] &\simeq h_1 a[n-1] + h_0 a[n]. \end{aligned} \tag{3}$$

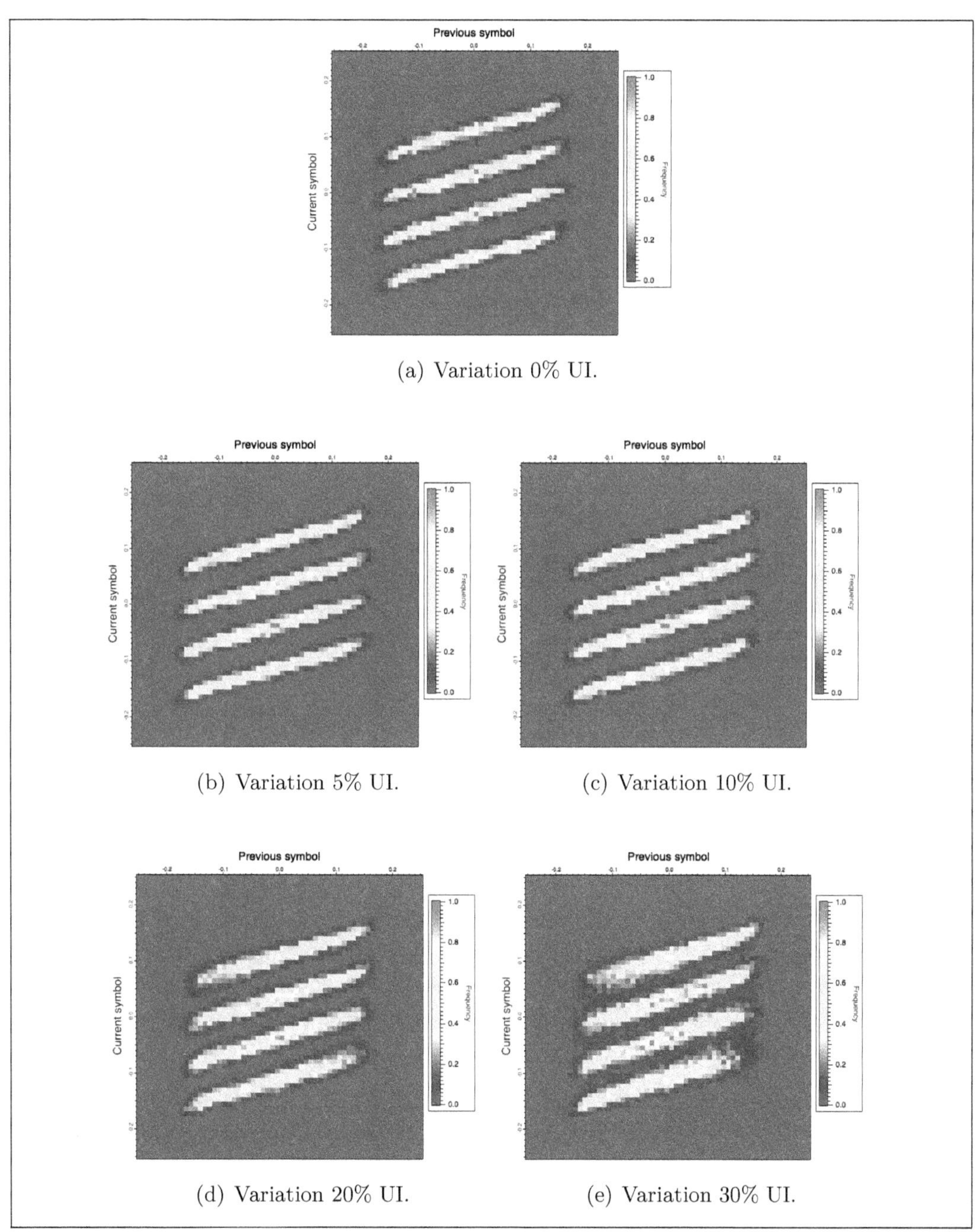

(a) Variation 0% UI.

(b) Variation 5% UI.

(c) Variation 10% UI.

(d) Variation 20% UI.

(e) Variation 30% UI.

Figure 12: 2D symbol mapping with variation of sampling at 4 Gbps measured PAM-4 signaling on the 1-m MSL (1 UI = 0.5 nsec).

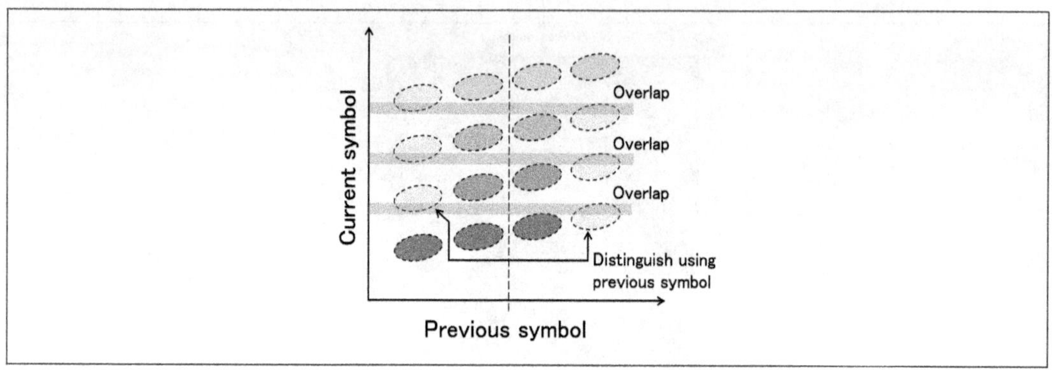

Figure 13: 2D symbol mapping with an overlapping adjacent symbol.

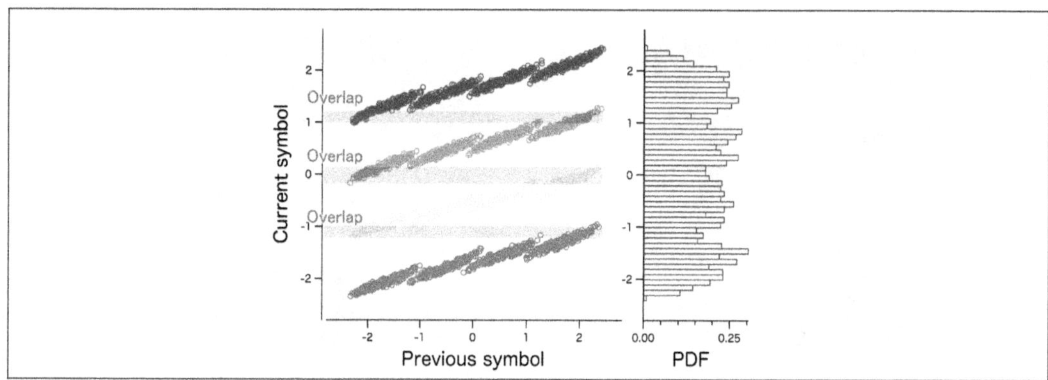

Figure 14: Simulation result of 2D mapping at 2 Gbps PAM-4 on the 2-m MSL.

By examining the term $h_1 a[n-1]$, we observe that as the value of $a[n-1]$ increases, $y[n]$ also increases. Similarly, from the term $h_{-1}a[n]$, we can observe that as $a[n]$ increases, $y[n-1]$ also increases. Hence, the coefficients h_1 and h_{-1}, representing pre-/post-cursor effects, can be interpreted as factors causing the 2D mapping to exhibit skewness, as depicted in Fig. 15.

Based on these relationships, the distributions of symbols, Δa_i and Δb_{i-j}, expand and follow the increasing symbol rate. The distributions of the same symbols align and combine with each other, resulting in Fig. 16(a). As a consequence, d_{i-j} approaches zero when $d_{i-j} < 0$, leading to a closed eye. In a 1D plot, each symbol distribution curve overlaps, making it challenging to discriminate individual symbols. However, in the 2D mapping, the symbol areas do not overlap, allowing for discrimination of each symbol, unlike in the 1D plot where overlapping occurs.

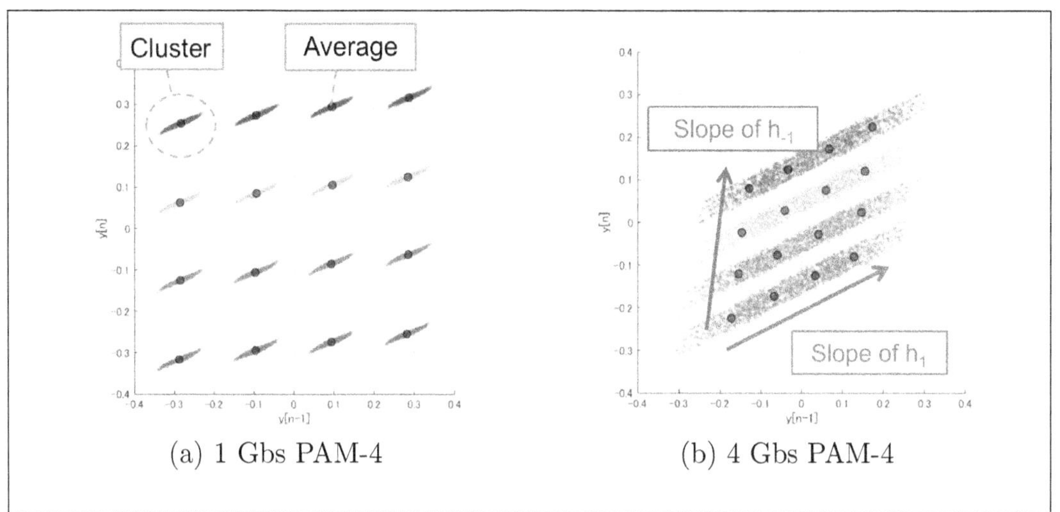

(a) 1 Gbs PAM-4

(b) 4 Gbs PAM-4

Figure 15: symbol clusters property of 2D mapping.

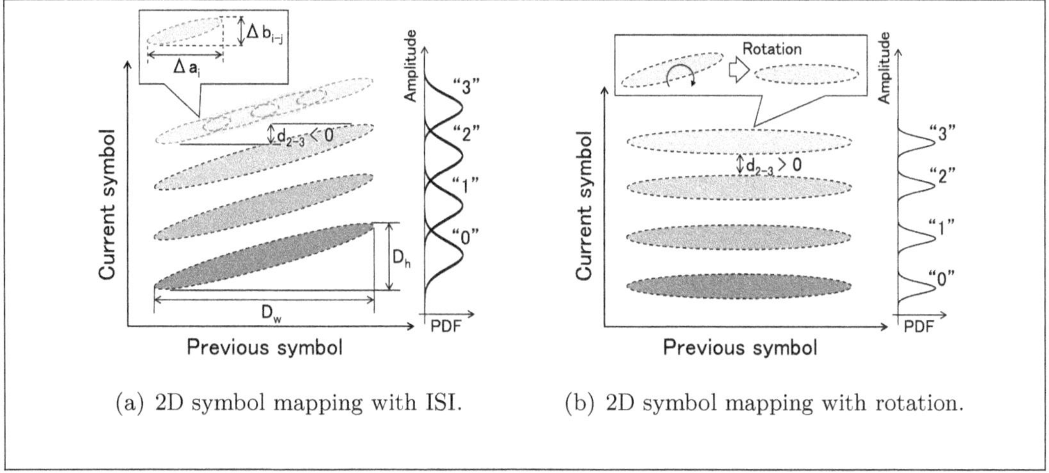

(a) 2D symbol mapping with ISI.

(b) 2D symbol mapping with rotation.

Figure 16: Symbol distribution of closed eye owing to ISI effect.

5.2 Visualization of equalizer effect using 2D mapping

As symbol rates increase, the spacing along the y-axis tends to reduce. Nevertheless, there can be situations where despite the reduced y-axis spacing in the 2D mapping, the spacing viewed from an oblique angle remains open. In these circumstances, if the slope can be adjusted such that the oblique spacing is rotated to align with the y-axis spacing, the eye diagram could present an open eye, as shown in Fig. 16.

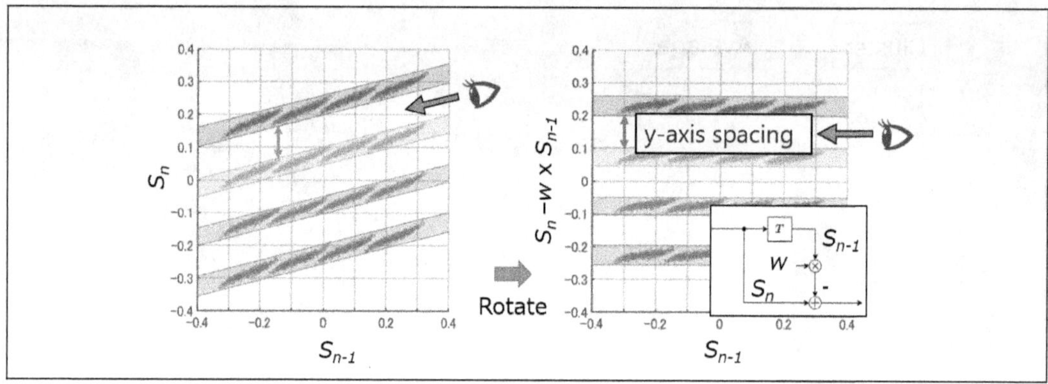

Figure 17: Symbol map rotation of received data.

Figure 16(b) illustrates the compensation of angles for each symbol distribution, resulting in symbol distances denoted as d_{i-j} becoming $d_{i-j} > 0$. To achieve this, the current symbol s_n is updated by rotating the distributions using the previous symbol s_{n-1}, according to the following relationship:

$$s_n' = s_n - w \cdot s_{n-1}. \tag{4}$$

The appropriate weight w can be calculated as $w = \frac{D_h}{D_w}$, where D_h and D_w are represented as depicted in Fig. 16(a). By rotating the symbols on a 2D surface, the overlap of each symbol distribution can be eliminated, and these symbol distributions are obtained by projecting the 2D mapped symbols onto the y-axis, as shown in Fig. 16(b). The block diagram of the symbol rotation at the receiver end is illustrated in Fig. 17. The outputs of the symbol rotation operation are achieved by subtracting the current symbol from the previous (delayed) symbol, as depicted in Fig. 17. The symbol rotation operation is equivalent to a 1st-order FFE. Symbol detection can be accomplished through conventional threshold discrimination by rotating the 2D mapping onto a 1D signal. This relationship demonstrates that the 2D mapping can visualize the equalizer effect, as shown in Fig. 18.

Figure 19 illustrates the simulation results of a transmission at 2.5 Gbps PAM-4 on a 2-m MSL, along with the 2D plot and symbol distribution resulting from the rotation process with different weights. The symbol distribution undergoes rotation as the weights are adjusted, resulting in vision changes in the symbol distribution. While the histograms of each symbol may overlap, as shown in Fig. 19(a), the shapes of the histograms are altered through symbol rotation, as depicted in Figs 19(b)-(d). When the weight is set to -0.3, the symbol distribution is rotated almost horizontally, separating each symbol from the distribution. The received symbols

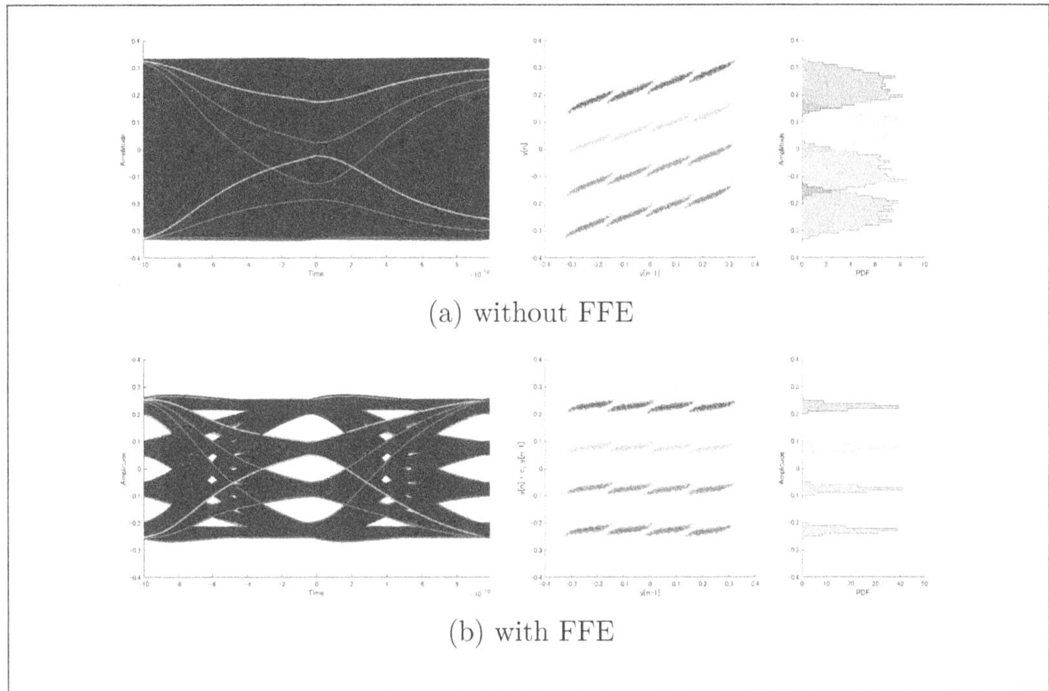

(a) without FFE

(b) with FFE

Figure 18: Visualization of FFE impact using 2D mapping.

can be accurately obtained by appropriately setting threshold levels for symbol discrimination.

Figure 20 displays a 2D plot depicting the results of the rotation process with various weights for the measured symbol data at the receiver end on a 1-m MSL at 4 Gbps. The symbol distribution undergoes rotation by adjusting the weights (corresponding to the FFE operation), making the distribution nearly horizontal when the weight is set to -0.3.

The utilization of 2D mapping for signal transmission quality evaluation represents a novel approach, and conducting a quantitative assessment compared to conventional methods presents a challenge for future research. However, in evaluating commonly encountered scenarios with challenging complete eye closure conditions, the 2D mapping technique provides qualitative advantages by enabling the assessment of symbol overlap. Furthermore, 2D mapping facilitates the visualization and evaluation of noise effects. Further theoretical investigations are planned to expand upon these findings.

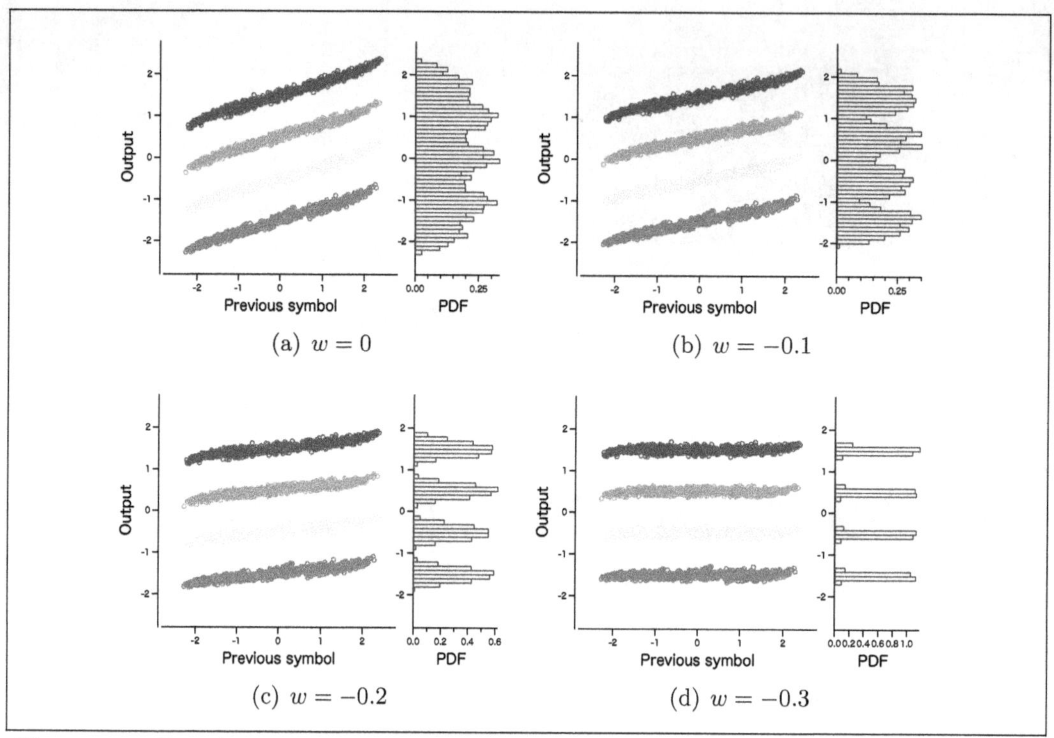

Figure 19: Simulation results of 2D symbol mapping with symbol rotation at 2.5 Gbps PAM-4 on the 2-m MSL (Sampling variation is 10% with 2,982 sampling points).

6 Conclusion

This paper proposed a novel technique for evaluating the quality of PAM-4 data transmission by employing a 2D symbol mapping method. This study advanced the current approach by integrating further theoretical elements, effectively expanding on the prior work proposed in [14] The 2D symbol mapping method enables the visualization of ISI, providing insights into the extent of its impact. The results obtained from simulations and experiments demonstrate that the data transition diagram effectively evaluates the quality of PAM-4 data transmission, even in the presence of degraded performance due to ISI. Furthermore, it allows for the visualization of the equalization effect.

This study primarily examines the visualization of post-cursor ISI influence using 2D mapping. However, future research efforts will expand this approach to incorporate 3D mapping, considering the pre-cursor effect.

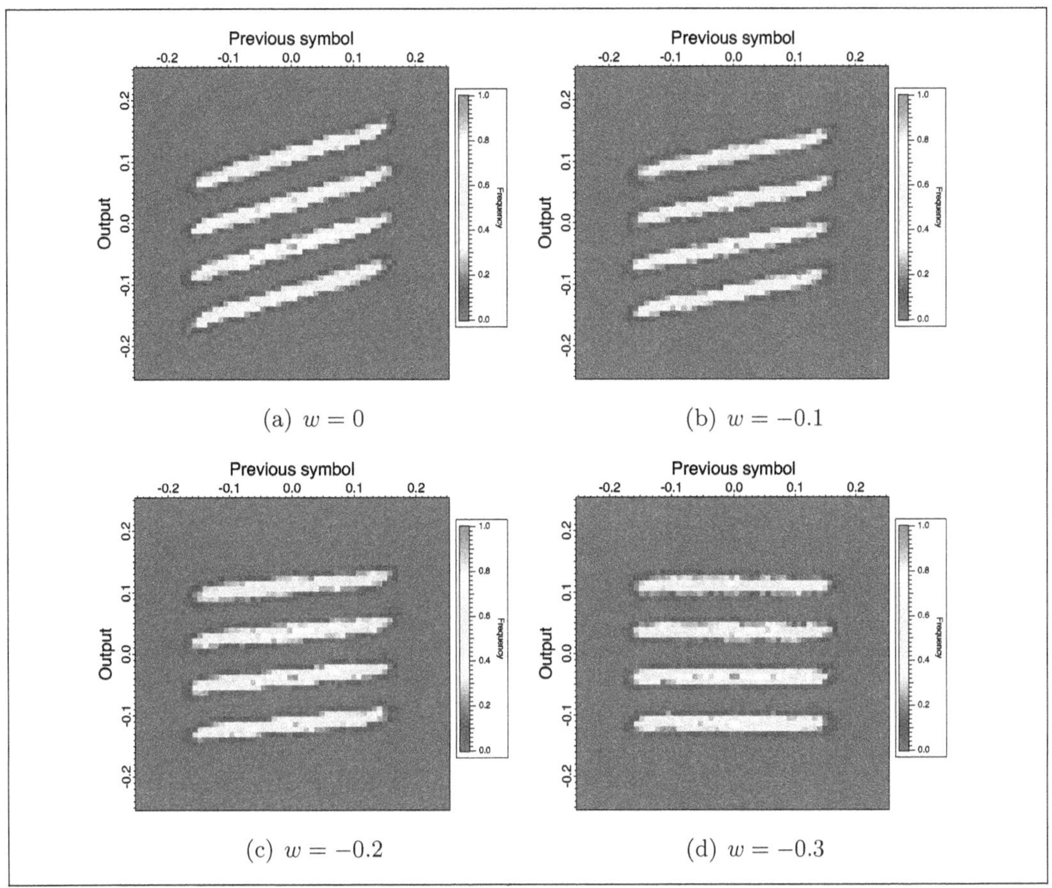

(a) $w = 0$ (b) $w = -0.1$

(c) $w = -0.2$ (d) $w = -0.3$

Figure 20: 2D symbol mapping of rotated symbol using measured results of received symbol at 4 Gbps PAM-4 on the 1-m MSL (Sampling variation 10% UI).

References

[1] IEEE P802.3bs 200 Gb/s and 400 Gb/s Ethernet Task Force, `http://www.ieee802.org/3/bs/`

[2] J. Barry, E. Lee, and D. Messerschmitt, Digital Communication, Springer, 2004.

[3] D. Cui, H. Zhang, N. Huang, A. Nazemi, B. Catli, H. G. Rhew, B. Zhang, A. Momtaz, and J. Cao, "3.2 A 320mW 32Gb/s 8b ADC-based PAM-4 Analog Front-End with Programmable Gain Control and Analog Peaking in 28nm CMOS," *2016 IEEE International Solid-State Circuits Conference (ISSCC)*, pp.58-59, 2016.

[4] G. Steffan, E. Depaoli, E. Monaco, N. Sabatino, W. Audoglio, A.A. Rossi, S. Erba, M. Bassi, and A. Mazzanti, "A 64Gb/s PAM-4 Transmitter with 4-Tap FFE and 2.26pJ/b Energy Efficiency in 28nm CMOS FDSOI," *2017 IEEE International Solid-State Cir-*

cuits Conference, pp.116–117, 2017.

[5] T.O. Dickson, H.A. Ainspan, and M. Meghelli, "A 1.8pJ/b 56Gb/s PAM-4 Transmitter with Fractionally Spaced FFE in 14nm CMOS," *2017 IEEE International Solid-State Circuits Conference*, pp.118–119, 2017.

[6] Y. Iijima and Y. Yuminaka, "Double-Rate Tomlinson-Harashima Precoding for Multi-Valued Data Transmission," *IEICE Transactions on Information & Systems*, vol.100-D, no.8, pp.1611–1617, 2017.

[7] Y. Yuminaka, T. Kitamura, and Y. lijima, "PAM-4 Eye Diagram Analysis and Its Monitoring Technique for Adaptive Pre-Emphasis for Multi-Valued Data Transmissions," *Proc. IEEE 47th International Symposium on Multiple-Valued Logic*, pp. 13–18, 2017.

[8] B. Analui, A. Rylyakov, S. Rylov, M. Meghelli, and A. Hajimiri, "A 10-Gb/s Two-Dimensional Eye-Opening Monitor in 0.13-μm Standard CMOS," *IEEE Journal of Solid-State Circuits*, vol.40, no.12, pp.2689–2699, 2005.

[9] Y. Iijima, K. Taya, and Y. Yuminaka, "PAM-4 Eye-Opening Monitoring Techniques using Gaussian Mixture Model," *IEEE 50th International Symposium on Multiple-Valued Logic*, pp.149–154, 2020.

[10] Y. Yuminaka, K. Taya, and Y. Iijima, "PAM-4 Eye-Opening Monitoring Techniques based on Gaussian Mixture Model Fitting," *IEICE Communication Express*, Vol.9, No.10, pp.464–469, 2020.

[11] K. Taya, Y. Yuminaka, and Y. Iijima, "Statistical Waveform Evaluation Method for Adaptive PAM-4 Equalization," *Journal of Applied Logics - IfCoLog Journal of Logics and their Applications*, Vol.8, No.4, pp.1087–1099, 2021.

[12] Y. Iijima and Y. Yuminaka, "PAM-4 Eye-Opening Monitor Technique Using Gaussian Mixture Model for Adaptive Equalization," *IEICE Transactions on Information & Systems*, vol.104-D, no.8, pp.1138–1145, 2021.

[13] Y. Iijima and Y. Yuminaka, "Efficient PAM-4 Data Transmission with Closed Eye Using Symbol Distribution Estimation," *IEEE 51st International Symposium on Multiple-Valued Logic*, pp.195–200, 2021.

[14] Y. Iijima K. Nakajima, and Y. Yuminaka, "Two-Dimensional Symbol Mapping for Evaluating Multi-Valued Data Transmission Quality," *IEEE 52nd International Symposium on Multiple-Valued Logic*, pp.170–175, 2022.

[15] F. Lu, P. Peng, S. Liu, M. Xu, S. Shen, and G. Chang, "Integration of Multivariate Gaussian Mixture Model for Enhanced PAM-4 Decoding Employing Basis Expansion," *2018 Optical Fiber Communications Conference and Exposition (OFC)*, pp.1–3, 2018.

[16] L. Sun, J. Du, J. Liu, B. Chen, K. Xu, B. Liu, C. Lu, and Z. He, "Intelligent 2-Dimensional Soft Decision Enabled by k-means Clustering for VCSEL-based 112-Gbps PAM-4 and PAM-8 Optical Interconnection," *Journal of Lightwave Technology*, vol.37, no.24, pp.6133–6146, 2019.

Quantum Algorithms for Unate and Binate Covering Problems with Application to Finite State Machine Minimization

Abdirahman Alasow
Portland State University, Portland, OR, USA
alasow@pdx.edu

Marek Perkowski
Portland State University, Portland, OR, USA
mperkows@ee.pdx.edu

Abstract

Covering problems find applications in many areas of computer science and engineering, such that numerous combinatorial problems can be formulated as covering problems. Combinatorial optimization problems are generally NP-hard problems that require an extensive search to find the optimal solution. Exploiting the benefits of quantum computing, we present a quantum oracle design for covering problems, taking advantage of Grover's search algorithm to achieve quadratic speedup. This paper also discusses applications of the quantum counter in unate covering problems and binate covering problems with some important practical applications, such as finding prime implicants of a Boolean function, implication graphs, and minimization of incompletely specified Finite State Machines.

1 Introduction

Many optimization problems can be formulated as the selection of a subset from a larger set. One familiar form is the covering problems [8, 20, 26]. Covering problems are minimization problems for combinatorial optimization problems in which a certain combinatorial structure covers another. For instance, in logic design, the covering problems can be formulated as a set of minterms to be covered by a set of subsets of minterms (such as the prime implicants for a specific example). Minterms for a Boolean function of n variables are products of literals for all variables of this function. A literal is a variable or the negation of a variable. True minterms are those for which the value of the function is 1. The prime implicant is the product of literals that cannot be extended by removing some of the literals and which covers some subset of true minterms. Covering problems are generally given as a table with rows corresponding to the set elements and columns corresponding to the subsets.

Covering problems include two main types: unate covering problem (UCP) and binate covering problem (BCP) [28, 12, 18]. The unate covering problem is the problem of finding a minimum cost assignment to variables to which a given Boolean function f is equal to 1. The literals are all in positive form (uncomplemented). The Binate covering problem has the extra constraint that negative literals may be present. The covering problem can be generalized by assuming that the choice of a subset implies the choice of another subset. This additional constraint can be represented by an implication clause. For example, if the selection of group a implies the choice of b ($a \Rightarrow b$), then the clause $\bar{a} + b$ is added. Note that the implication clause makes the product of sums form binate in variable a, because a (uncomplemented) is also part of some covering clauses. Therefore, this class of problems is called binate covering or covering with closure. There are two main ways to express the covering problems: constraints in the form of a matrix or a product of sum (POS) form of a Boolean equation $f = 1$. This formulation of POS is known as Petrick's method [33]. If the covering problems are expressed in the form of a matrix, then the corresponding matrix (covering matrix) of the UCP is filled with elements from the set $\{1, 0\}$ while the BCP is filled with elements from the set $\{1, -1, 0\}$. A -1 entry corresponds to a complemented variable, and a 1 entry to an uncomplemented one. In the covering matrix, the rows correspond to each term of the expression, and the columns correspond to each variable. Covering problems can also be expressed in the form of a POS formula, such that $f(x_1, x_2, x_3, x_4) = (x_1 + x_2)(x_2 + x_3)(x_1 + x_4)x_3x_4 = 1$ then this is called an unate covering, which always has a solution. If some of the variables in the function appear both positive and negative (complement) such that $f(x_1, x_2, x_3, x_4) = (x_1 + x_3 + x_4)(x_1 + \overline{x_2} + x_4)(\overline{x_2} + x_3 + x_4)(x_2 + \overline{x_3} + x_4)$ then it is called binate covering, which may or may not have a solution. In this paper, we will express the covering problems in POS form.

2 Related Work

The fundamental covering problem known from computer science has many practical applications in electronic design automation (EDA) and digital systems.

2.1 Applications of Covering Problems

There are many combinatorial optimization problems for covering problems in various areas, such as logic minimization [35], scheduling [27], parallel computation on a GPU [30], and allocation, encoding, and routing [7] for unate covering problem. Binate covering problem is used in finite state machine minimization [12, 17], technology mapping [28], Boolean relations [15], and Directed Acyclic Graphs (DAG) covering problems [35, 24, 25]. Logic minimization is the process of finding the optimal implementation of a logic function. The minimum-cost implementation is achieved when the cover of a given function consists of the minimum number of prime implicants to represent the function. In another variant, the solution should have the total minimum cost of selected prime implicants. The Unate covering can be used to achieve exact logic minimization [35]. Scheduling problems can be formulated as UCP, such as vehicle and crew scheduling problems in [27], to minimize the combined vehicle and crew cost. Parallel computation on a GPU [30, 40, 39, 38] was explored using the Unate covering problem. The classical task of the GPU is matrix multiplication, which can be mapped to the UCP implementation. Also, UCP has many other applications in logistic problems such as allocation, encoding, and routing [7]. Finite state machine minimization [12, 17] can be formulated as a binate covering problem such that the number of internal states in the finite state machine can be reduced. A standard-cell library must first comply with the available library primitives in VLSI, a process known as technology mapping. Finding the optimal mapping of logic gates to VLSI library cells can be accomplished via binate covering [28]. Boolean relations such as two-level logic minimization under the Sum-of-Product (SOP) representation can be solved based on the binate covering problem formulation [15]. Finding the minimum set of nodes or paths covering every node in a directed acyclic graph is called DAG covering. This can be formulated as a binate covering problem [35, 24, 25] by constructing the closure condition of the graph.

2.2 Classical Algorithms for Covering Problems

The covering problem is considered NP-hard [34, 16], and much effort has been spent on it because of its wide applications. Several classical algorithms are proposed for covering problems based on exact and heuristic algorithms. Several exact algorithms are proposed for covering problems, such as the most widely known approach, the branch and bound algorithm [41, 15, 36, 22], with many techniques suggested for lower bound and upper bound improvement using pruning techniques. In the branch and bound technique, the covering table is expressed as the POS of the constraints, and the problem solution explores, in the worst case, all possible solution instances. Branch and bound employs the upper bound and lower bound methods. For each solution to the constraints, upper bounds on the value of the cost function are identified, and lower bounds are estimated using the current set of variable assignments. The upper bound value is updated each time a new lower-cost solution is discovered. When the lower bound estimation is greater than or equal to the most recently computed upper bound, the search can be pruned. As a result, a better solution will not be found using the current variable assignments, allowing us to prune the search. Branch and bound used reduction and bounding techniques of search space for solutions to avoid the generation of some of the suboptimal solutions. The upper bound is the cost of the cheapest solution seen so far. Eventually, it will be the cost of the optimal solution. While the lower bound is an estimate of the minimum cost of a solution for the problem. The lower bound is the most important factor for runtime.

A Binary Decision Diagrams (BDDs) based algorithm [41] was proposed to solve the covering problem such that finding the solution only requires computing the shortest path in the BDD. The number of variables in the BDD is equal to the number of columns in the binate table. However, the BDD tree is too large to be built when there are many variables. A mixed technique of both branch-bound and BDD-based algorithms was proposed in [15], such that the constraints are represented as a conjunction of BDDs. This method leads to an effective method to compute a lower bound on the cost of the solution. Exact algorithms are computationally expensive for large problems because of exhaustive searches. Although there are several improvements using pruning and reduction techniques for exact algorithms, in general, the computational time can be exponential. Thus, a heuristic approach [16, 37] has been proposed for covering problems that provide suboptimal solutions. The heuristic approach in [37] has time complexity $O(n^2m)$ where n is the number of variables and m the number of terms in the covering problem. While the literature proposes both exact and heuristic approaches for classical algorithms, there is still a need for efficient algorithms that may take advantage of quantum algorithms to solve covering problems efficiently.

3 Quantum Algorithm for Covering Problem

Since classical optimization techniques are inefficient for solving NP-hard problems in terms of computational complexity, we present a quantum algorithm for solving the covering problems. To the best of our knowledge, this is the first quantum algorithm for solving covering problems. Algorithms with quantum oracles are better than the corresponding classical search algorithms because they operate using quantum parallelism and quantum superposition, with all vectors being potential solutions simultaneously (all minterms). Thus, a quantum oracle that iterates sufficiently many times highly increases the probability of finding one of the solutions in a single measurement of all input qubits. Grover's algorithm implemented in quantum circuits gives a quadratic speedup when compared to an exhaustive classical algorithm for the same problem. Our paper reduces all covering problems to Grover's algorithm with an innovative way of building the quantum oracle that allows the number of qubits to be reduced logarithmically and at the same time solves both the decision and optimization problems in various variants.

A hybrid algorithm (a combination of classical and quantum) can be used to solve the covering problems, which assumes an arbitrary number of terms and variables. This algorithm would be a direct generalization of the algorithm presented in this paper. A classical computer can use any type of heuristic or algorithmic search method to expand a search tree. The upper part of the tree is created on a classical computer using all kinds of general search strategies, heuristic functions, cost functions, parameters, and constraints such as those discussed in [32, 18, 22]. This way, the sizes of the macro-leaves of the tree are reduced step by step such that the number of terms in them is less than m and the number of variables is less than n. For each macro-leaf, a quantum computer is called and executes a full search based on Grover's algorithm. Suppose the problem in the macro-leaves with less than m terms and less than n variables is recognized as a special type of problem. In that case, a special algorithm on a classical computer is executed for this reduced tree. This is a standard method used recently by several authors because Grover's algorithm gives a quadratic speedup only for problems for which a more efficient algorithm than a complete search does not exist. Also, this method allows for excellent scalability by sharing subtasks between the classical and quantum processors based on the availability of the quantum computer size (parameters of m and n).

3.1 Grover's Search Algorithm

Grover's algorithm [29, 42], searches an unordered array of N elements to find a particular element with a given property. In classical computations, in the worst case, this search takes N queries (tests and evaluations of the classical oracle). In the average case, a particular element will be found in $\frac{N}{2}$ queries. Grover's algorithm can find elements in $\sqrt{N}$ queries. Thus, Grover's algorithm can be used to find all possible solutions for covering problems. Grover's algorithm is a quantum search algorithm that speeds up a classical search algorithm of complexity $O(N)$ to $O(\sqrt{N})$ in the space of N objects. Hence, Grover's algorithm gives a quadratic speedup. To solve all optimal solutions of the covering problem, Grover's algorithm has to be repeated.

The covering problem contains n variables from the given Boolean function that are used to represent the search space of $N = 2^n$ elements. To solve the covering problem in Grover's algorithm, these N elements are applied in a superposition state, which is the input to the oracle. If the oracle recognizes an element as the solution, then the phase of the desired state is inverted. This is called the phase inversion of the marked element. The marked element is a true minterm of function f from the oracle. The true minterm is a product of all variables of function f that evaluates to $f = 1$. Grover's search algorithm uses another trick called inversion about the mean (average), also known as diffusion operation or amplitude amplification. Inversion about the mean amplifies the amplitude of the marked states and shrinks the amplitudes of other items. The amplitude amplification increases the probability of measuring the marked states, so that measuring the final states will return the target solution with a high probability near to 1.

An oracle is a black box operation that takes an input and gives an output, a yes/no decision. A quantum oracle is realized as a binary reversible circuit that is used in quantum algorithms for the estimation of the value of the Boolean function realized in it. The quantum oracle also has to replicate all input variables on the respective output qubits. If the function of the oracle is not reversible, we use ancilla qubits to make the function reversible. If the oracle uses ancilla qubits initialized to $|0\rangle$, it has to return also a $|0\rangle$ for every ancilla qubit. The classical oracle function is defined as a Boolean function $f(x)$ which takes a proposed solution x of the search problem. If x is the solution, then $f(x) = 1$; If x is not a solution, then $f(x) = 0$. The quantum oracle is a unitary operator O such that:

$$|x\rangle|y\rangle \rightarrow |x\rangle|q \oplus f(x)\rangle$$

where x is the value in search space, q is a single qubit, the oracle qubit, and $\oplus$ is the EXOR operator (also called the addition modulo 2). A simplified formula of the quantum oracle can be written as:

$$|x\rangle \rightarrow (-1)^{f(x)}|x\rangle$$

As shown in Figure 1a, the n qubits in the superposition state result from applying a vector of Hadamard gates to the initial state $|0\rangle^n$. Next, we applied a repeated operator G which is called the Grover's Loop. After the iteration of the Grover's Loop operator $O(\sqrt{N})$ times the output is measured for all input qubits. Oracle can use an arbitrary number of ancilla qubits, but all these qubits must be returned to value $|0\rangle$ inside the oracle. The Grover's Loop G is a quantum subroutine that can be broken into four steps, as shown in Figure 1b:

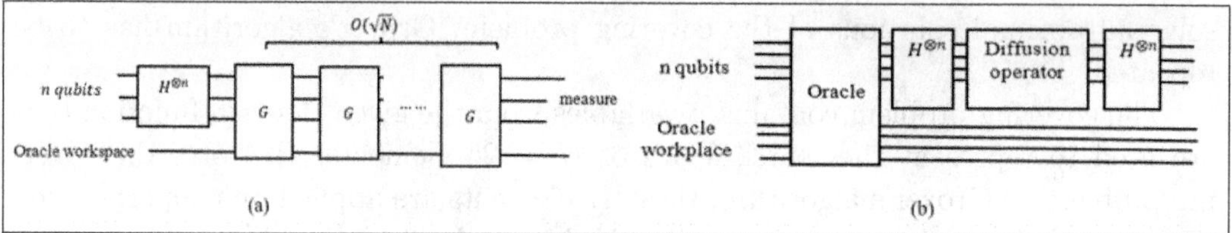

Figure 1: (a) Schematic circuit for Grover's algorithm [29]. (b) Grover's Loop Operator

1. Apply the Oracle O. This step is phase inversion.
2. Apply the Hadamard transform $H^{\otimes n}$, where $H = \frac{1}{\sqrt{2}}\begin{bmatrix} 1 & 1 \\ 1 & -1 \end{bmatrix}$
3. Perform the condition phase shift also known as a zero state phase shift, in which all states receive a phase shift of -1 except for the zero state $|0\rangle$. This step is also known as the diffusion operator.
4. Apply the Hadamard transform $H^{\otimes n}$

The number of required iterations for Grover's operator is: $R \leq \left\lceil \frac{\pi}{4}\sqrt{\frac{N}{M}} \right\rceil$ where M is number of solutions and N is number of all search space elements. The literature includes many methods for efficient and optimal design of circuits in quantum oracle [6, 13, 19, 11].

3.2 Quantum Counter

Each term of the clause required one multi-input Toffoli gate that has n qubit for the clause literals and one extra ancilla qubit to store the result of the clause. This was the traditional design of the quantum oracle, where each clause would have one ancilla qubit. For large problems, the number of ancilla qubits is very large, so even future quantum computers could not handle this approach. We propose an advanced quantum oracle design based on a quantum counter [3] that logarithmically reduces the number of required qubits. The quantum counter block is built from multi-input Toffoli and CNOT gates, where the first qubit of the quantum counter is applied as a constant 1 with the other qubits together.

In Figure 2a z is the least significant qubit and x the most significant. The outputs of CNOT and two of the Toffoli gates are $1 \oplus z, 1 \cdot z \oplus y$, and $1 \cdot z \cdot y \oplus x$ respectively. When $xyz = 000$, the first Toffoli gate outputs $1 \cdot z \cdot y \oplus x = 1 \cdot 0 \cdot 0 \oplus 0 = 0 \oplus 0 = 0$ and the second $1 \cdot z \oplus y = 1 \cdot 0 \oplus 0 = 0 \oplus 0 = 0$. The outputs of the qubits

y and x are both zeros. The output of the qubit z is $1 \oplus z = 1 \oplus 0 = 1$. Hence the circuit incremented 000 by 1 to 001. Quantum counter circuit indeed outputs the value $input + 1$.

If we connect the first control input of the quantum counter block to a circuit, then the output of the connected circuit (a term of the POS) will either activate or deactivate the counter. When the output of the connected circuit is equal to 1, the output of the counter block is incremented by 1. When the output of the circuit is equal to 0, the output of the counter block is unchanged. Below is a table in Figure 2b for all the input combinations of the 3-qubit quantum counter.

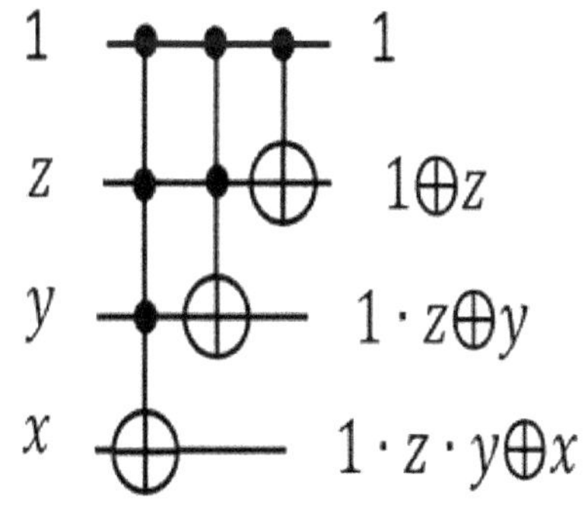

(a)

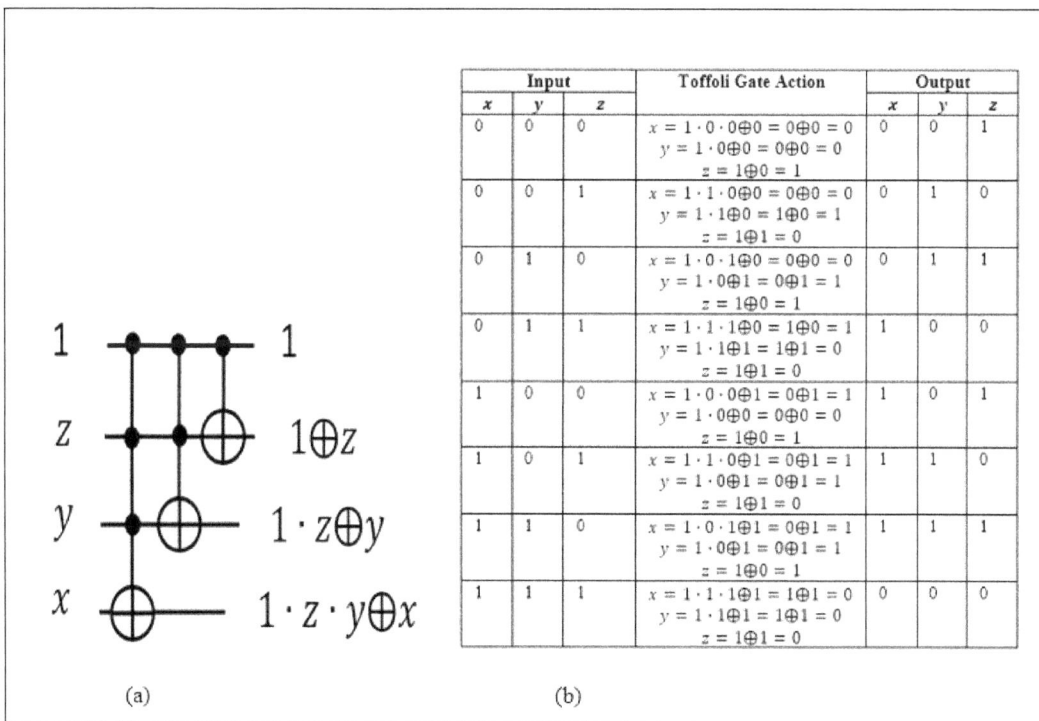

Input			Toffoli Gate Action	Output		
x	y	z		x	y	z
0	0	0	$x = 1\cdot 0\cdot 0\oplus 0 = 0\oplus 0 = 0$ $y = 1\cdot 0\oplus 0 = 0\oplus 0 = 0$ $z = 1\oplus 0 = 1$	0	0	1
0	0	1	$x = 1\cdot 1\cdot 0\oplus 0 = 0\oplus 0 = 0$ $y = 1\cdot 1\oplus 0 = 1\oplus 0 = 1$ $z = 1\oplus 1 = 0$	0	1	0
0	1	0	$x = 1\cdot 0\cdot 1\oplus 0 = 0\oplus 0 = 0$ $y = 1\cdot 0\oplus 1 = 0\oplus 1 = 1$ $z = 1\oplus 0 = 1$	0	1	1
0	1	1	$x = 1\cdot 1\cdot 1\oplus 0 = 1\oplus 0 = 1$ $y = 1\cdot 1\oplus 1 = 1\oplus 1 = 0$ $z = 1\oplus 1 = 0$	1	0	0
1	0	0	$x = 1\cdot 0\cdot 0\oplus 1 = 0\oplus 1 = 1$ $y = 1\cdot 0\oplus 0 = 0\oplus 0 = 0$ $z = 1\oplus 0 = 1$	1	0	1
1	0	1	$x = 1\cdot 1\cdot 0\oplus 1 = 0\oplus 1 = 1$ $y = 1\cdot 0\oplus 1 = 0\oplus 1 = 1$ $z = 1\oplus 1 = 0$	1	1	0
1	1	0	$x = 1\cdot 0\cdot 1\oplus 1 = 0\oplus 1 = 1$ $y = 1\cdot 0\oplus 1 = 0\oplus 1 = 1$ $z = 1\oplus 0 = 1$	1	1	1
1	1	1	$x = 1\cdot 1\cdot 1\oplus 1 = 1\oplus 1 = 0$ $y = 1\cdot 1\oplus 1 = 1\oplus 1 = 0$ $z = 1\oplus 1 = 0$	0	0	0

(b)

Figure 2: (a) Three-qubit quantum counter. (b) Analysis of a 3-qubit quantum counter block from (a)

We assign a counter block for each OR term (clause) and individual terms in the Boolean function for the covering problem, where the result of the clause is used as one of the control qubits of the counter. When the clause evaluates to 0, the counter forwards its input to the output without change. When it evaluates to 1, the counter outputs the binary number $value + 1$ to the previously accumulated count $value$. The use of a quantum counter allows us to send the result from the Toffoli gate representing one OR term to the counter circuit, hence eliminating the need for an ancilla qubit. We can set the function qubit back to 1 by mirroring the Toffoli gate used to compute the result and set the input qubits back to the original by applying NOT gates when appropriate. Our design drastically reduces the number of qubits needed for a function at the cost of replicating Toffoli gates in the POS expression and the costs of the iterative counter.

Our proposed concept of using a quantum counter can be used to design a quantum oracle for both decision and optimization problems, such as SAT-like, MAX-SAT problems, and many other problems in machine learning, such as mining frequent pattern generation [4]. In traditional quantum oracle design for SAT-like, MAX-SAT, and covering problems, every clause is built as a multi-input Toffoli gate. The number of qubits in the multi-input Toffoli gate is equal to the number of variables in the clause plus one extra ancilla qubit to save the result of the clause. In our design, there is no need for extra ancilla qubits for each clause, but several clauses have a quantum counter which shares ancilla qubits. For instance, if there are 30 clauses, our design requires only 5 ancilla qubits for all 30 clauses, rather than the 30 ancilla qubits required in the traditional quantum oracle. Thus, our design reduces the number of qubits logarithmically.

4 Grover Algorithm for Unate Covering Problem

The unate covering problem (also called the set covering problem [21]) is to find the minimum cost assignment of variables that satisfy a Boolean equation $f = 1$. The literals in the POS function f are all in the positive form (uncomplemented variables).

4.1 Finding All Prime Implicants for the Exact Minimum Covering of a SOP Circuit

Minterms for a Boolean function of n variables are products of literals for all variables of the function. A literal is a variable or the negation of a variable. True minterms are those for which the value of the function is 1. The prime implicant in a sum-of-product (SOP) structure is a product of literals that cannot be extended by removing some of the literals and which covers some subset of true minterms. For instance, given a function from Figure 3a, all its prime implicants are marked as ovals (loops). Using the minterm compatibility graph G, all prime implicants are found as maximum cliques. Prime implicants can also be found as maximum independent sets of graphs (G complement). Based on the truth table (or a Karnaugh Map) and the prime implicants, the covering table from Figure 3b is created. In this table, every row represents a prime implicant as a product of Boolean literals, and each column represents a true minterm. The covering table is filled with symbol X for every minterm included in a prime implicant.

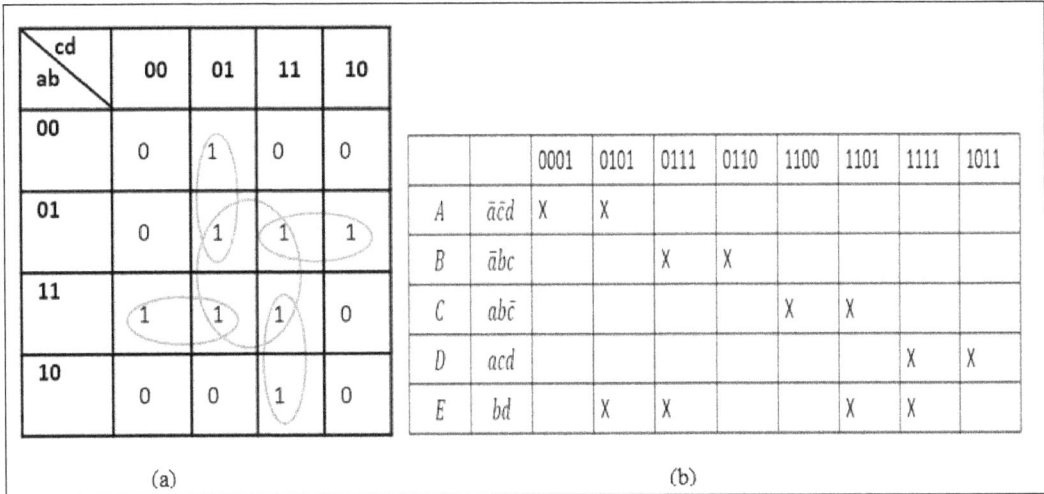

cd \ ab	00	01	11	10
00	0	1	0	0
01	0	1	1	1
11	1	1	1	0
10	0	0	1	0

(a)

		0001	0101	0111	0110	1100	1101	1111	1011
A	$\bar{a}\bar{c}d$	X	X						
B	$\bar{a}bc$			X	X				
C	$ab\bar{c}$					X	X		
D	acd							X	X
E	bd		X	X			X	X	

(b)

Figure 3: (a) Truth table for all prime implicants of a SOP circuit. (b) Covering table for SOP function from (a)

From the table in Figure 3b, denoting rows A, B, C, D, E we compile the Petrick function in a standard way such that each column is created as one term by adding the variables in the row corresponding with symbol X. For instance, column 0101 has two cells filled with X. Adding the two rows $A+E$ as one term that corresponds to the symbols X in column 0101. In such a way, equation (1) is created:

$$A(A+E)(B+E)BD(D+E)(C+E)C = 1 \tag{1}$$

Equation (1) can be simplified using the Boolean law: $A(A+E) = A \cdot A + A \cdot E = A(1+E) = A$ to the following equation: $1 = A \cdot B \cdot C \cdot D$. Therefore, $f = A + B + C + D = \bar{a}\bar{c}d + \bar{a}bc + acd + ab\bar{c}$ is the minimum sum of products of expression of function f. In the case of many variables and clauses, solving this problem exactly is very difficult.

The main goal of this example is to show how SOP minimization can be solved using the UCP. Then the quantum oracle is built from UCP to apply Grover's algorithm. The quantum oracle is built from UCP (1). First, we need to convert each OR term into a product using De Morgan's Law $A+E = \overline{\overline{A+E}} = \overline{\overline{A}\,\overline{E}} = 1 \oplus \overline{A}\,\overline{E}$; $B+E = \overline{\overline{B+E}} = \overline{\overline{B}\,\overline{E}} = 1 \oplus \overline{B}\,\overline{E}$; $C+E = \overline{\overline{C+E}} = \overline{\overline{C}\,\overline{E}} = 1 \oplus \overline{C}\,\overline{E}$; $D+E = \overline{\overline{D+E}} = \overline{\overline{D}\,\overline{E}} = 1 \oplus \overline{D}\,\overline{E}$. From (1) we can rearrange $A \cdot B \cdot C \cdot D(A+E)(B+E)(D+E)(C+E)$.

To simplify the oracle design, we consider $A \cdot B \cdot C \cdot D$ as one term, which needs one quantum circuit. Then we connect one block of the iterative quantum counter after each Toffoli gate representing the OR term of the function POS formula. We put the ancilla qubit back to its original state by mirroring each Toffoli gate after each counter. In this case, we need five quantum counter circuits, as can be seen in Figure 4. Also, we need to add the NOT gate in the output circuit block, which makes the last qubit out_0 to produce 1 if the variable xyz in counter circuit is 101 such that the Boolean function in equation (1) is equal to 1.

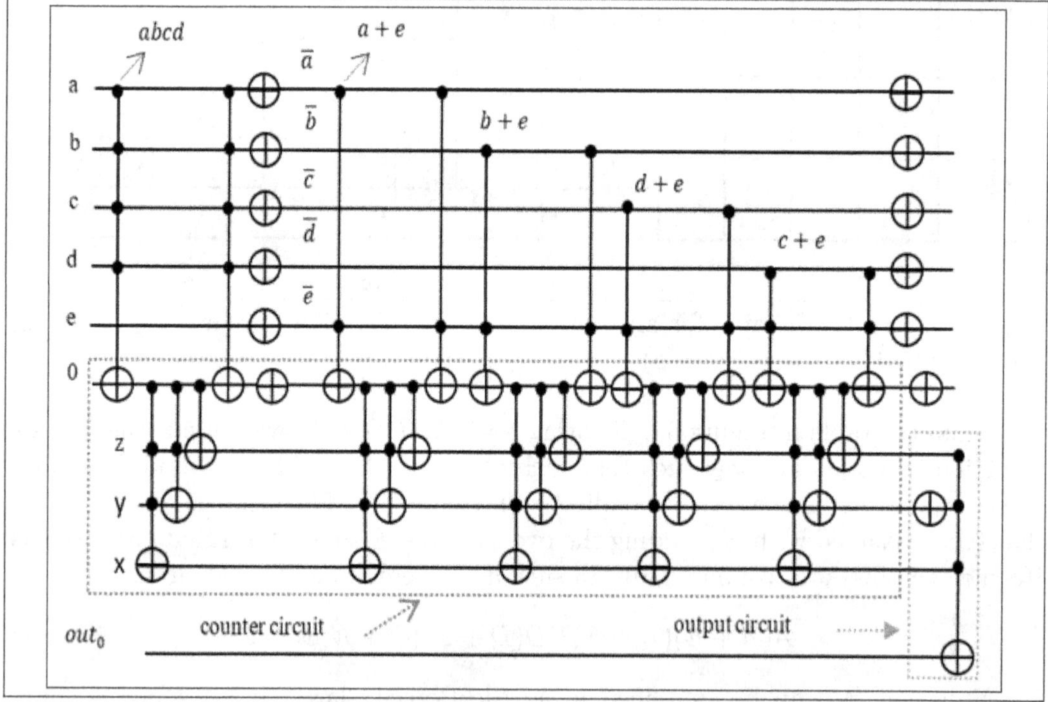

Figure 4: Quantum oracle for $A \cdot B \cdot C \cdot D(A+E)(B+E)(D+E)(C+E)$

In Figure 5, we applied the oracle circuit in Figure 4 in the Grover's search algorithm for iterations $R = 4$ from this formula: $R \leq \left\lceil \frac{\pi}{4}\sqrt{\frac{N}{M}} \right\rceil$ where $N = 2^5 = 32$ is the number of all search space elements since there are 5 variables for A, B, C, D, E. $M = 2$ is the number of solutions. $M = 2$ because in equation (1) is equal to 1 either $ABCDE = 11110$ or $ABCDE = 11111$. We run the circuit on the 'qasm_simulator' from QISKIT for 1024 shots (independent runs to get high precision probability) which the circuit produces the correct answers. We measured a_0, a_1, a_2, a_3 and a_4 in Figure 5 where a_0, a_1, a_2, a_3, a_4 correspond to the Boolean variables A, B, C, D, E re-

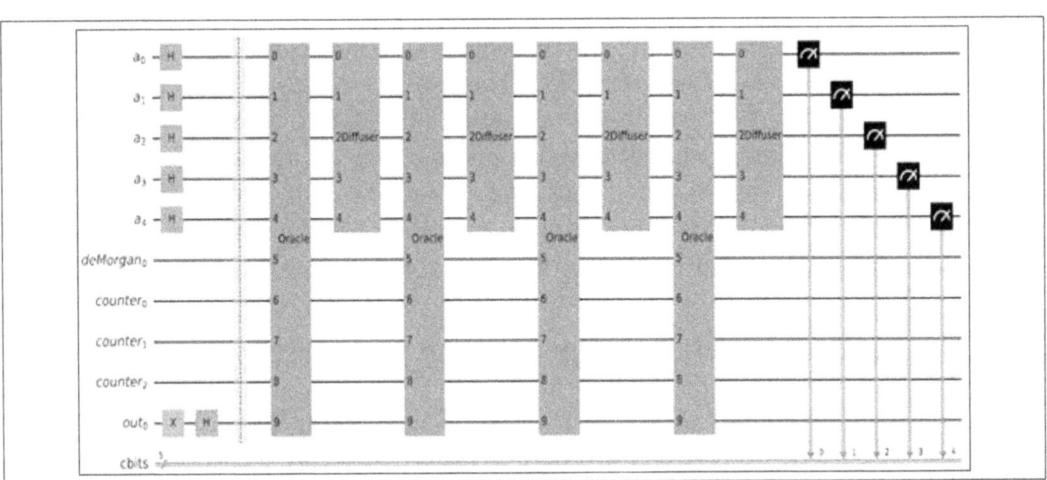

Figure 5: Grover's algorithm with 4 iterations using the oracle circuit from Figure 4

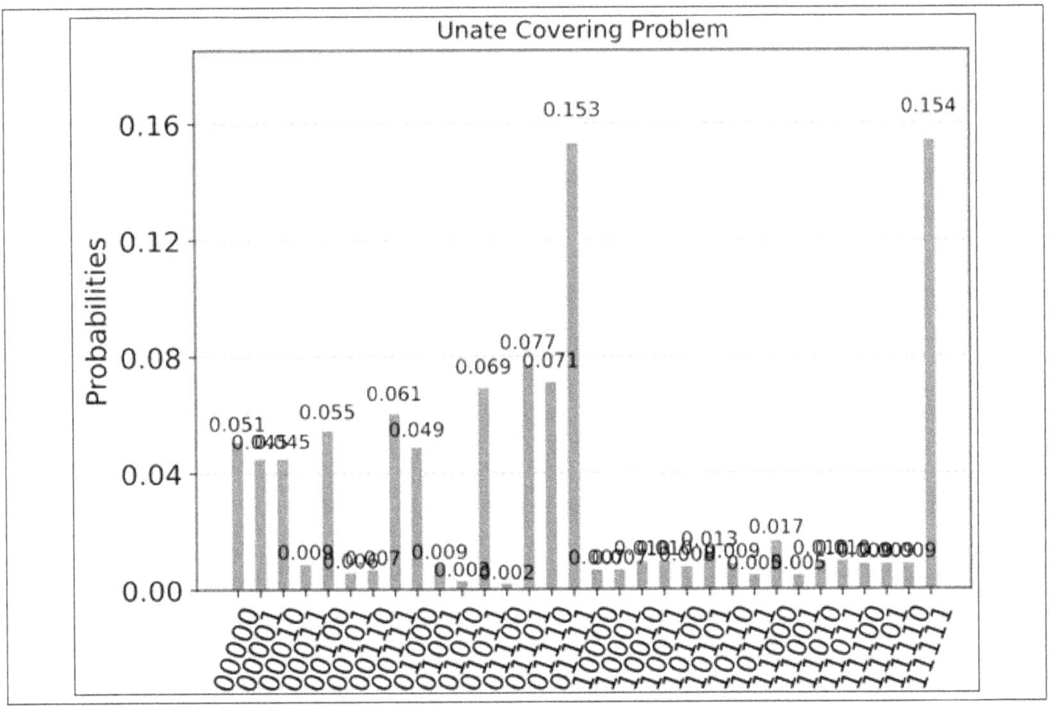

Figure 6: Measurement of the Boolean variables from $A \cdot B \cdot C \cdot D(A + E)(B + E)(D + E)(C + E)$

spectively. As can be seen in Figure 6, the diagram illustrates the QISKIT [5] output graphics for the simulated circuit. The measured values $a_4a_3a_2a_1a_0$ with high probability are 11111, 01111. The values 11111 and 01111 correspond to E, D, C, B, A respectively which are two solutions to equation (1): A, B, C, D and A, B, C, D, E.

This example explains that the above method can be applied to an arbitrary POS formula with x variables and y clauses. Therefore, our method is scalable to an arbitrary size of unate covering problem, assuming a sufficiently large number of qubits in a quantum computer or in a quantum component of a hybrid computer.

5 Grover's Algorithm for Binate Covering Problem

Binate covering problem is the same as unate covering problem with the additional constraint that BCP can contain negative literals. There are some cases that have no solution, and therefore, the BCP should reliably notify the user about this fact. We presented three examples of BCP problems: (1) Finding the minimum covering for an implication graph. (2) Finding a minimum cost constraint. (3) Minimization of incompletely specified Finite State Machines.

5.1 Finding the Minimum Covering for an Implication Graph

An implication graph is a directed acyclic graph (DAG) where each node represents a variable assignment. An implication graph represents the implication relations between pairs of variable assignments. Given a set S and a family of subsets $F = \{s_1, s_2, \cdots, s_n\}$, a closure conditions are represented as an implication graph. For instance, assuming given a family of subsets of the set $\{1, 2, 3, 4, 5, 6\}$ and the implication graph from in Figure 7, the optimization task is to select a subset of nodes from the set A, B, C, D, E that satisfies all three conditions:

- Covering condition: all items from the set $\{1, 2, 3, 4, 5, 6\}$ must be covered by the selected nodes.
- All closure conditions must be satisfied.
- The set of selected nodes must have the minimum number of elements.

The general optimization can be solved by constructing a covering and closure table. For a particular problem, as specified above, the table is shown in Figure 8. Here, the rows correspond to the nodes (subsets) of the implication graph in Figure 7, while the columns correspond to the individual elements of the set $\{1, 2, 3, 4, 5, 6\}$ for covering table and the columns for the closure table correspond to the nodes in the implication graph which are the same subsets as the rows of the table.

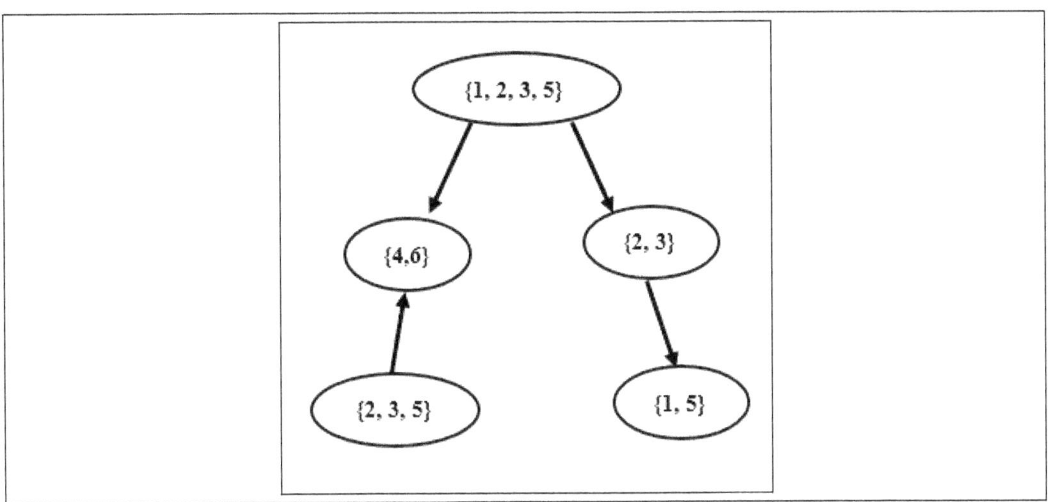

Figure 7: Implication graph for the set $\{1, 2, 3, 4, 5, 6\}$

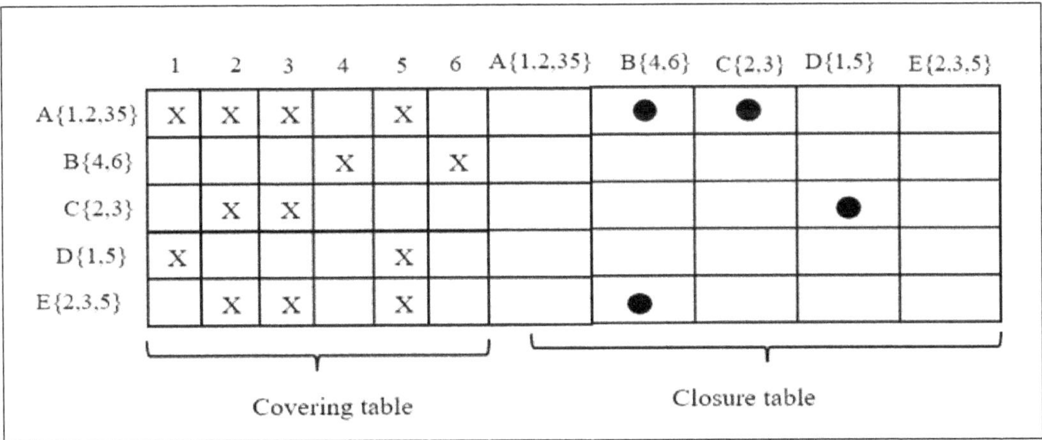

	1	2	3	4	5	6	A{1.2.35}	B{4.6}	C{2.3}	D{1.5}	E{2.3.5}
A{1.2.35}	X	X	X		X			●	●		
B{4.6}				X		X					
C{2.3}		X	X							●	
D{1.5}	X				X						
E{2.3.5}		X	X		X			●			

Figure 8: Covering-closure table based on the implication graph from Figure 7

The closure conditions are illustrated in Figure 7. Assuming that we select set $A = \{1, 2, 3, 5\}$ on top of the implication graph, it is implied that the set $B = \{4, 6\}$ and the set $C = \{2, 3\}$ must be also selected. Set $\{2, 3\}$ implies set $\{1, 5\}$ and set $\{2, 3, 5\}$ implies set $\{4, 6\}$ (see Figure 7 above). This way, the table from Figure 8 is created. For instance, the black dots in the row A mean that set $A = \{1, 2, 3, 5\}$ implies sets $B = \{4, 6\}$ and $C = \{2, 3\}$. Set $C = \{2, 3\}$ implies sets $D = \{1, 5\}$. Set $E = \{2, 3, 5\}$ implies sets $B = \{4, 6\}$. Based on the covering and closure table from Figure 8, Petrick's method creates a Boolean formula in equation (2). Next, this formula is transformed into the general POS formula from equation (3)

$$(A+D)(A+C+E)B(A+D+E)(A \Rightarrow BC)(C \Rightarrow D)(E \Rightarrow B) = 1 \tag{2}$$

In equation (2), the first four terms describe the covering conditions, and the last three terms correspond to closure conditions (closure constraints). Equation (2) can be converted to equation (3) by using the logic transformation rule $(A \Rightarrow B) \Leftrightarrow (\overline{A} + B)$

$$(A+D)(A+C+E)B(A+D+E)(\overline{A}+BC)(\overline{C}+D)(\overline{E}+B) \tag{3}$$

Next , we build a quantum oracle for the general POS formula from equation (3) by converting each OR term into Products using De Morgan's law $A+D = \overline{\overline{A+D}} = \overline{\overline{A}\,\overline{D}} = 1 \oplus \overline{A}\,\overline{D}$; $A+C+E = \overline{\overline{A+C+E}} = \overline{\overline{A}\,\overline{C}\,\overline{E}} = 1 \oplus \overline{A}\,\overline{C}\,\overline{E}$; $A+D+E = \overline{\overline{A+D+E}} = \overline{\overline{A}\,\overline{D}\,\overline{E}} = 1 \oplus \overline{A}\,\overline{D}\,\overline{E}$; $\overline{A}+BC = \overline{\overline{\overline{A}+BC}} = \overline{A\overline{BC}} = 1 \oplus A\overline{BC} = 1 \oplus A(1 \oplus BC)$; $\overline{C}+D = \overline{\overline{\overline{C}+D}} = \overline{C\overline{D}} = 1 \oplus C\overline{D}$; $\overline{E}+B = \overline{\overline{\overline{E}+B}} = \overline{E\overline{B}} = 1 \oplus E\overline{B}$. Then we connect one block of the iterative quantum counter after each Toffoli gate, representing the OR term of the function POS formula. We put the ancilla qubit back to its original state by mirroring each Toffoli gate after each counter. In this case, we need seven quantum counter circuits (one for B term and six for each POS term), as can be seen in Figure 9. The oracle circuit in Figure 9 is inserted into Grover's Oracle. This is similar to previous examples that demonstrate one more time how inefficient using the global AND is, even for very small practical binate covering problems. Rather than using AND, we use a quantum counter.

This example illustrates that an arbitrary size problem of solving an implication graph can be reduced to algorithmically creating Grover's oracle with x variable qubits and y terms. Therefore, the implication graph problem can be solved by Grover's algorithm with a quadratic speedup.

In Figure 10, we applied the oracle circuit from Figure 9 in Grover's search algorithm for iterations $R = 2$ from this formula: $R \leq \left\lceil \frac{\pi}{4}\sqrt{\frac{N}{M}} \right\rceil$ where $N = 2^5 = 32$ and $M = 5$ (this can be verified by a truth table). We run the circuit on the 'qasm_simulator' from QISKIT for 1024 shots, and the circuit produces the correct answers. We measured a_0, a_1, a_2, a_3 and a_4 in Figure 10 where a_0, a_1, a_2, a_3, a_4 corresponds to the Boolean variables A, B, C, D, E respectively. As can be seen in the histogram in Figure 11, this illustrates the QISKIT [5] output graphics for the simulated circuit. The measured values $a_4a_3a_2a_1a_0$ with high probability are 01110, 01111, 11010, 11110, and 11111 corresponding to variables E, D, C, B, A respectively. For instance, the vector 01110 corresponds to the solution of DCB.

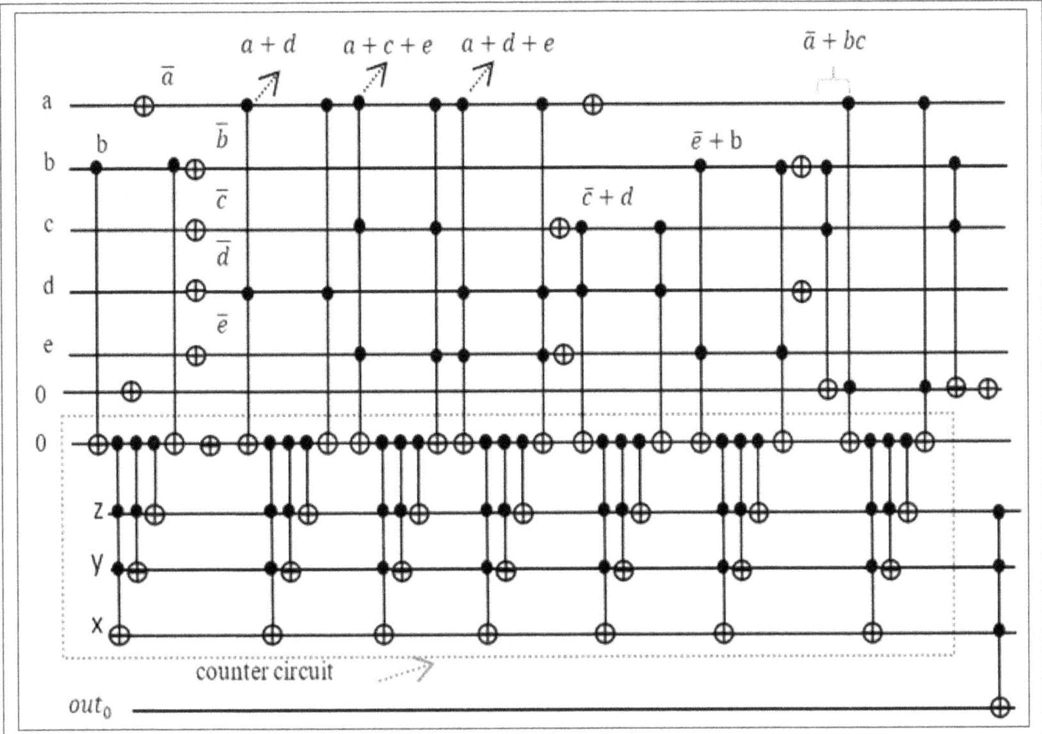

Figure 9: Quantum oracle for $(A+D)(A+C+E)B(A+D+E)(\overline{A}+BC)(\overline{C}+D)(\overline{E}+B)$

In the measurement, the solutions have much higher probabilities than the non-solutions. These solutions are verified outside of Grover's algorithm, just by using the oracle with function $(A+D)(A+C+E)B(A+D+E)(\overline{A}+BC)(\overline{C}+D)(\overline{E}+B)$.

5.2 Finding Minimum Cost Constraint

Minimum cost constraint minimizes the cost of satisfying the assignment for the given problem. Given m constraints on n Boolean variables, the goal is to find an assignment that satisfies all constraints such that:

$$minimizing \sum_{i=0}^{n-1} w_i x_i$$

subject to $f = y_0 \wedge y_1 \wedge \cdots \wedge y_{m-1} = 1$.

Where $w_i \geq 0$ is the weight of variable x_i and y_m is a clause which means a sum of x_i. This is called a weighted binate covering. The practical application of finding

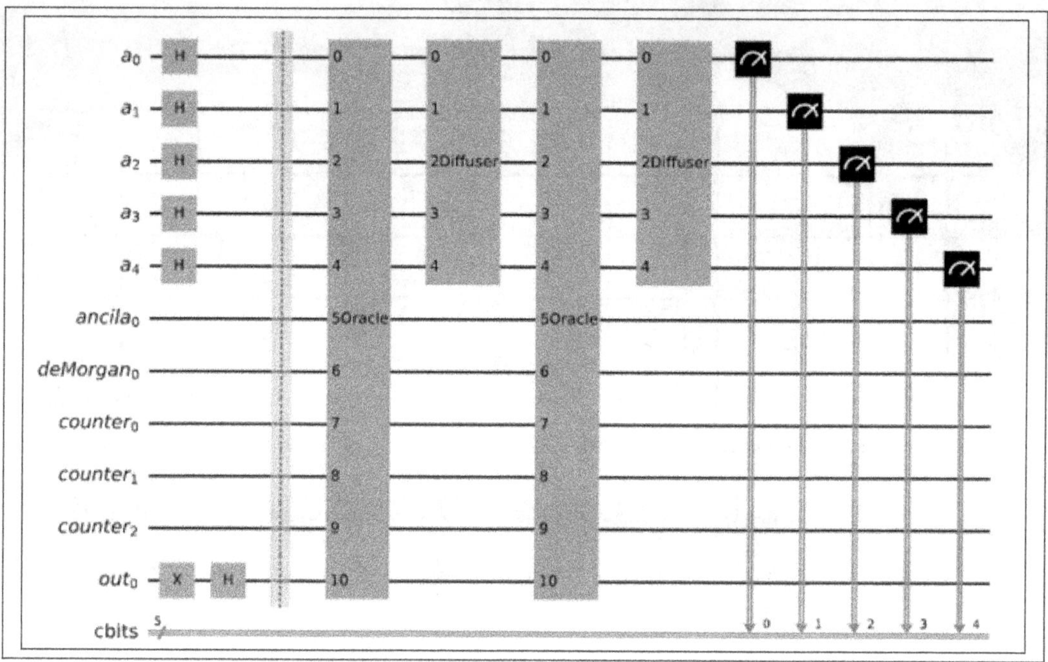

Figure 10: Grover's algorithm with 2 iterations using the oracle circuit

the minimum cost constraint can be found in [21], which can be formulated as a binate covering problem. However, we present in this paper a general example of a Boolean function that can be solved using Grover's algorithm. For instance, given is the following set of clauses Y:

$$y_0 = x_0 + x_3$$

$$y_1 = x_2 + \overline{x_3}$$

$$y_2 = x_1 + \overline{x_2}$$

$$y_3 = x_1 + x_2 + x_3$$

and let the weight w_i for each value of x_i as follow: $w_0 = 4, w_1 = 2, w_2 = 1$, and $w_3 = 1$. The optimization task is to find the minimum cost assignment based on the given weight. First, we construct the POS function, and then we design a quantum oracle for the Grover's algorithm to find the exact minimum solution. Based on the solution, we apply the weight for each clause to find the minimum cost constraint. Here the Y clauses are represented by a POS Boolean formula:

$$(x_0 + x_3)(x_2 + \overline{x_3})(x_1 + \overline{x_2})(x_1 + x_2 + x_3) \tag{4}$$

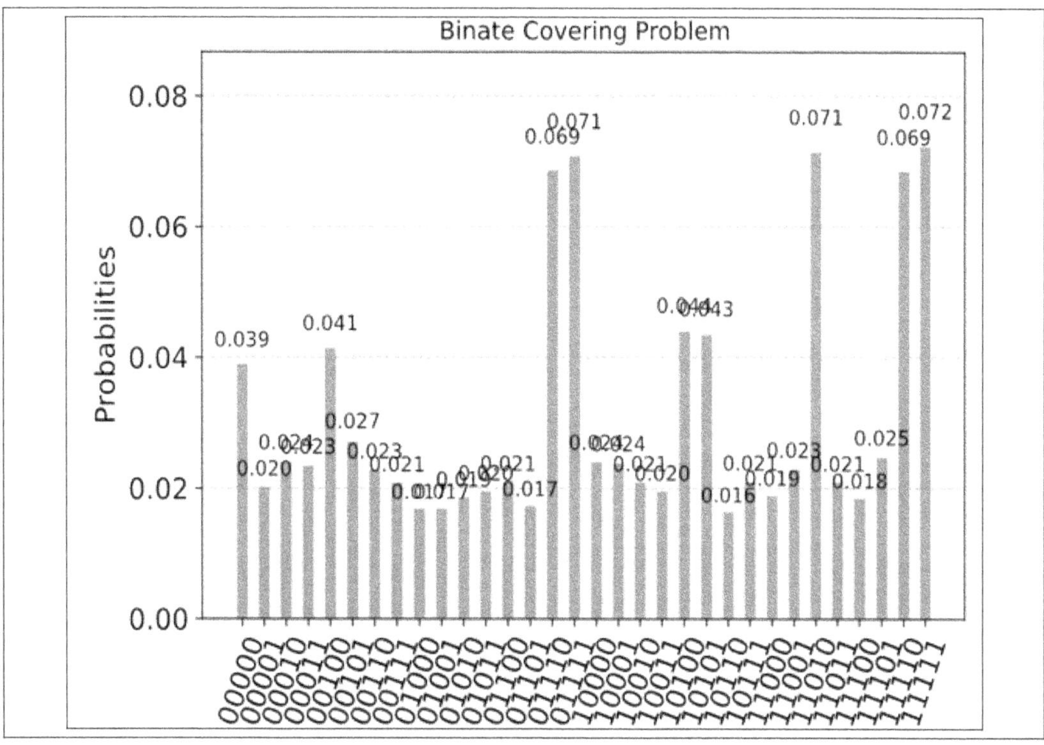

Figure 11: Measurement of the Boolean variables from Figure 10 based on $ABCDE$

The solution found in Grover's algorithm is subject to the equation (5) for the minimum cost function:

$$4x_0 + 2x_1 + x_2 + x_3 \tag{5}$$

Equation (5) is an arithmetic function that is computed in a classical processor of a hybrid computer, while equation (4) is computed in Grover's algorithm. First, we build a quantum oracle from equation (4), similar to the previous examples, after applying De Morgan's low: $x_0 + x_3 = \overline{x_0}\,\overline{x_3} \oplus 1$; $x_2 + \overline{x_3} = \overline{x_2}x_3 \oplus 1$; $x_1 + \overline{x_2} = \overline{x_1}x_2 \oplus 1$; $x_1 + x_2 + x_3 = \overline{x_1}\,\overline{x_2}\,\overline{x_3} \oplus 1$

We need to add the two NOT gates in the output circuit block, which makes the last qubit out_0 to produce one if the variable xyz in counter circuit is 100, such that the Boolean function in equation (4) is equal to 1.

In Figure 13, we applied the oracle circuit in Figure 12 in Grover's search algorithm for iterations R=2 from this formula: $R \leq \left\lceil \frac{\pi}{4}\sqrt{\frac{N}{M}} \right\rceil$ where $N = 2^4 = 16$ is the number of all search space elements. $M = 4$ is the number of solutions that can be verified by creating a truth table. We run the circuit on the 'qasm_simulator' from

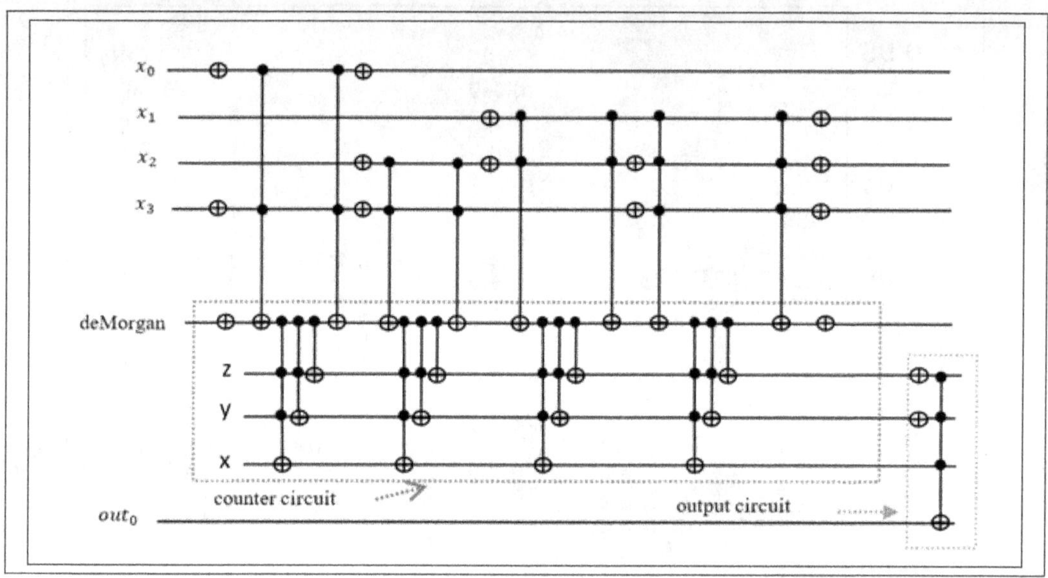

Figure 12: Quantum oracle for $(x_0 + x_3)(x_2 + \overline{x_3})(x_1 + \overline{x_2})(x_1 + x_2 + x_3)$

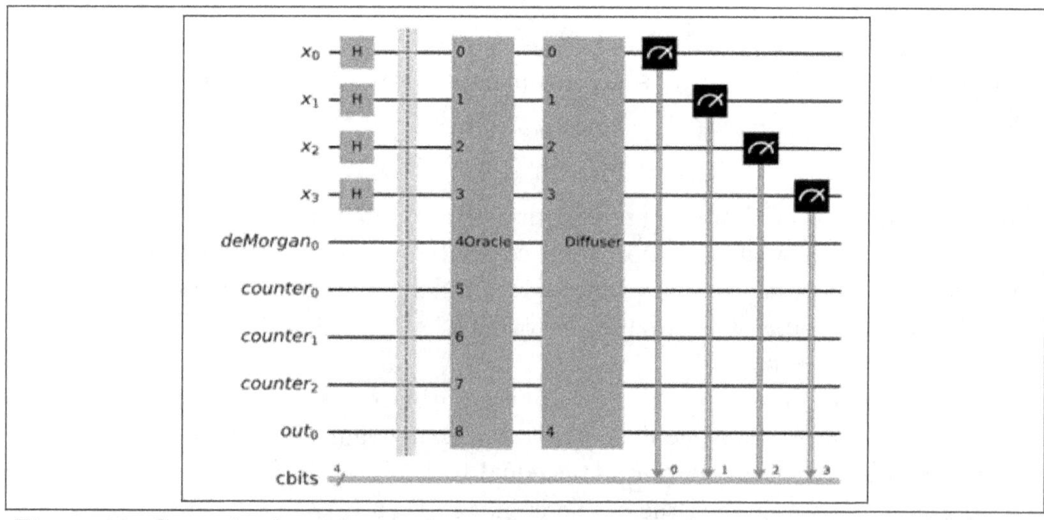

Figure 13: Grover's algorithm with 2 iterations using the oracle circuit from Figure 12

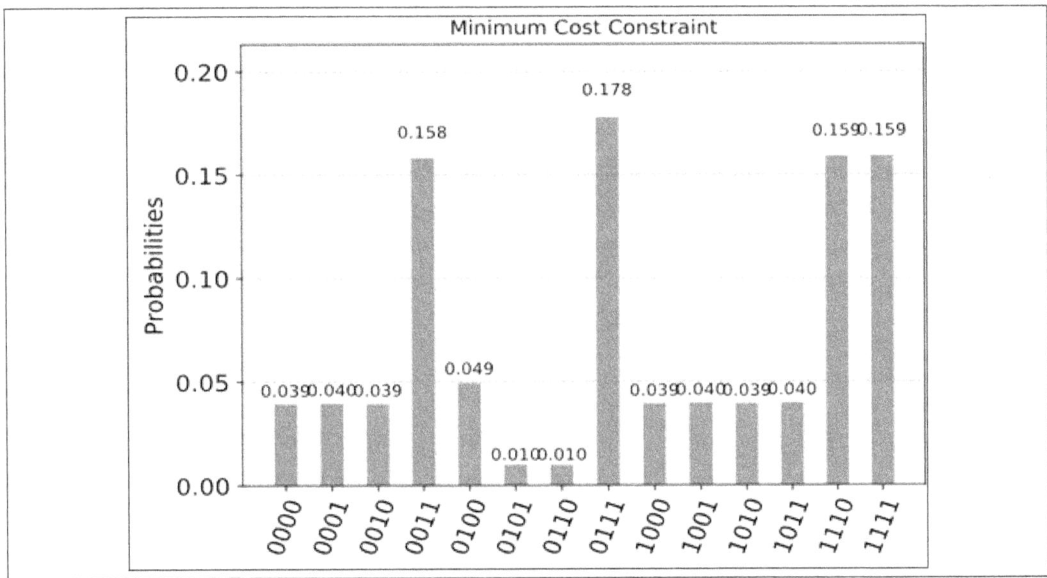

Figure 14: Measurement of the Boolean variables from Figure 13 based on $x_0x_1x_2x_3$

QISKIT for 1024 shots, and the circuit produces the correct answers. We measured $x_3x_2x_1x_0$ in Figure 13. As can be seen in Figure 14, the measured values $x_3x_2x_1x_0$ with high probability are {1111,1110,0111,0011}. Applying these values to equation (4), we found 1110, which gives the minimum cost of 4.

$$1110 : 4x_0 + 2x_1 + x_2 + x_3 = 4 * 0 + 2 * 1 + 1 * 1 + 1 * 1 = 4$$

Based on this value, we choose 1110 from the answer {1111,1110,0111,0011}, which corresponds to the solution of $x_3x_2x_1$ respectively with the minimum cost of 4. In another variant of our method, the arithmetic calculation is built into a quantum counter inside the oracle, such that for clause i instead of value 1, the value of w_i is added. This general method, however, is not practical for current quantum simulators. Concluding, the presented method is scalable to arbitrary size problem of minimizing the minimum cost constraint can be reduced to algorithmically creating Grover's oracle with x variable qubits and y terms. Therefore, finding the minimum cost constraint can be solved by Grover's algorithm with a quadratic speedup.

5.3 Minimization of Incompletely Specified Finite State Machines

A Finite State Machine (FSM) is an abstract model used in design to model problems in various fields of science and engineering. A finite state machine consists of input states, output states, and internal states that can change from one state to another state based on input. The change of the internal state is described by a transition function. For various reasons, finite state machines can have incompletely specified output functions and transition functions. The minimization of incompletely specified finite state machines is considered an NP-hard problem [17]. We present the complete oracle design for Grover's algorithm for the well-known classical problem of minimization of the number of states of incompletely specified finite state machines. For a given incompletely specified finite state machine, the solution is achieved by the following steps:

1. Classical computers create a triangular table to obtain compatible states.
2. Based on the triangular table, the classical computer creates a compatibility graph.
3. Quantum computer finds all maximum cliques in the compatibility graph.
4. The classical computers create the covering table component of the covering-closure based on the maximum cliques.
5. The classical computer creates a closure table component of the covering-closure table only for compatible states.
6. The classical computer creates a Boolean function for the oracle from the covering-closure table.
7. A quantum oracle is designed by a classical computer.
8. Grover's algorithm is called on a quantum computer with the oracle found in point 7.

Below we will illustrate the above general hybrid algorithm on a particular example. Given is an incompletely specified Mealy finite state machine described as a transition/output table from Figure 15a. Dashes represent don't care in internal states or output states. This table has internal states A, B, C, D, E, F, two input signals for columns, and one binary output under a slash symbol in cells of the map. A triangular table in Figure 15b was generated based on the table from Figure 15a. The table from Figure 15b covers all possible cases to minimize the number of states in the finite state machine. The 'X' symbol in the table indicates no possibility for grouping the corresponding states. Symbol 'V' in the table indicates that the states can be combined without any problem. A pair of state variables in a cell of the triangular map V indicates that states can be grouped only if the states mentioned in the block can be combined without any problem. For instance, states B and F can be combined under the condition that states C, F are compatible (can be combined). The method of creating the triangular table is well-known from [20].

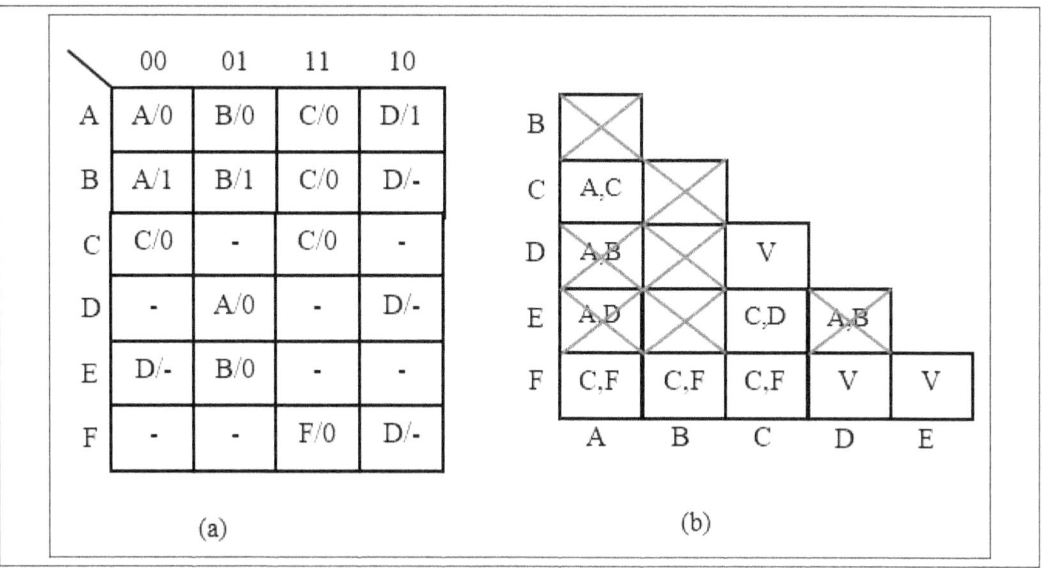

	00	01	11	10
A	A/0	B/0	C/0	D/1
B	A/1	B/1	C/0	D/-
C	C/0	-	C/0	-
D	-	A/0	-	D/-
E	D/-	B/0	-	-
F	-	-	F/0	D/-

(a)

Figure 15: (a) Truth table of FSM (b) FSM triangular table generated based on the truth table from (a)

Based on the triangular table, a compatibility graph for the state machine is generated, as shown in Figure 16a. Every cell in the triangular map that has no symbol X corresponds to a compatible pair of states. For instance, the cell at the intersection of row C and column A is compatible. The cell at the intersection of row B and column A is not compatible as it has a symbol X in it. Similarly, states A and D are not compatible. In Figure 15b, states A and D are compatible under the condition that states A and B are compatible. But we found earlier that states A and B are not compatible; thus, states A and D are not compatible. The cell at the intersection of row D and column A is crossed-out as symbol X. In the same way, the cell at the intersection of column A and row E is replaced with X because states A and D are not compatible. A simple recursive classical algorithm creates symbols X for every incompatible pair of internal states.

From Figure 16a, the maximum cliques of the graph are identified as $\{A, C, F\}$, $\{B, F\}$, $\{C, D, F\}$, $\{C, E, F\}$. Finding of all cliques in a graph is done by a SAT-based algorithm similar to those discussed earlier in this paper. The compatibility graph from Figure 16a is a complement of the incompatibility graph from Figure

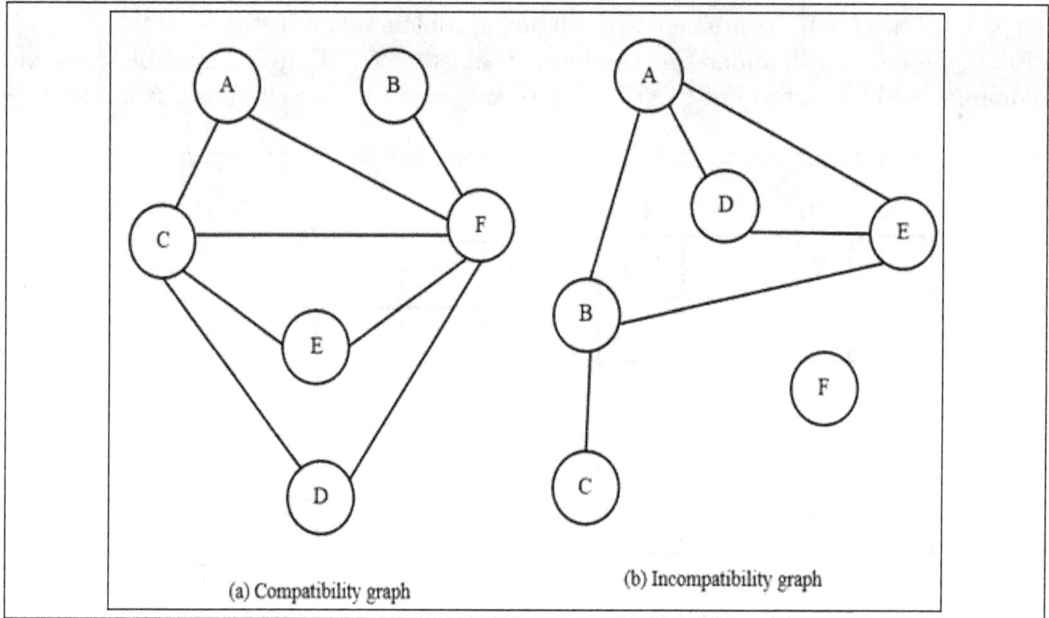

Figure 16: (a) Compatibility graph (b) Incompatibility graph.

16b. As we see, the maximum cliques in the compatibility graph are the same as the maximum independent sets in the incompatibility graph from Figure 16b. These maximum independent sets can be found from the graph coloring [14] of the incompatibility graph. The graph coloring can be solved by a special Grover's oracle. It can be solved by finding the Maximum Independent Sets and then solving the unate covering problem with them. These problems are reducible to SAT-like oracles for Grover's algorithm. This example explains the relation between the graph coloring of the incompatibility graph, finding the maximum cliques of the compatibility graph, and covering problems. These partial quantum algorithms are also useful in a quantum algorithm for solving the Ashenhurst-Curtis Decomposition [23].

To minimize the finite state machine, a covering-closure table, shown in Figure 17, is created by considering the maximum cliques and all their subsets as rows of the table. All of the states $\{A, B, C, D, E, F\}$ in the machine correspond to columns of the covering table, and the implications $\{A, C\}, \{C, F\}, \{C, D\}$ in the compatibility graph from the cell of the triangular table correspond to columns of the closure table. From the table in Figure 17, a binate covering problem can be specified using the equation:

$(X+V+P)Y(X+Z+U+V+Q+R+T)(Z+R+S)(U+T+W)(X+Y+$

	A	B	C	D	E	F	V{A,C}	Q{C,F}	R{C,D}
X{A,C,F}	X		X			X	●	●	
Y{B,F}		X				X		●	
Z{C,E,F}			X	X		X		●	
U{C,E,F}			X		X	X		●	●
V{A,C}	X		X				●		
P{A,F}	X					X		●	
Q{C,F}			X			X		●	
R{C,D}			X	X					
S{D,F}				X		X			
T{C,E}			X		X				●
W{E,F}					X	X			
	Covering table						Closure table		

Figure 17: Covering-Closure table for the FSM

$Z+U+P+Q+S+W)(X \Rightarrow VQ)$
$(Y \Rightarrow Q)(Z \Rightarrow Q)(U \Rightarrow QR)(V \Rightarrow V)(P \Rightarrow Q)(Q \Rightarrow Q)(T \Rightarrow R) = 1$

The function can be simplified using the Boolean laws $A \Rightarrow B \Leftrightarrow (\overline{A}+B)$.

$$\begin{cases}(X+V+P)Y(X+Z+U+V+Q+R+T)(Z+R+S)(U+T+\\ W)(X+Y+Z+U+P+Q+S+W)(\overline{X}+VQ)(\overline{Y}+Q)(\overline{Z}+Q)(\overline{U}+\\ QR)(\overline{V}+V)(\overline{P}+Q)(\overline{Q}+Q)(\overline{T}+R)=1\end{cases} \quad (6)$$

The number of search space $N = 2^{11} = 2048$ where 11 is the number of variables in the rows of covering-closure table. There are 155 solutions that the Boolean equation (6) is equal to 1. The presented method is not yet practical as contemporary quantum computers have not enough qubits. However, with a sufficient number of qubits, the presented algorithm will allow to minimize large machines with quadratic speedup. To visualize all these solutions in a histogram is difficult such that we use a more general case in Figure 18, we repeat the Grover's algorithm for iterations $R = 3$ with tuning values of thresholds until equal to counter value. The comparator $G = H$

compares the output from the counter with the threshold value given as constant values n_1, n_2, n_3 and n_4. (Using the threshold with a comparator has many other applications such as finding the minimum set of support [31]).

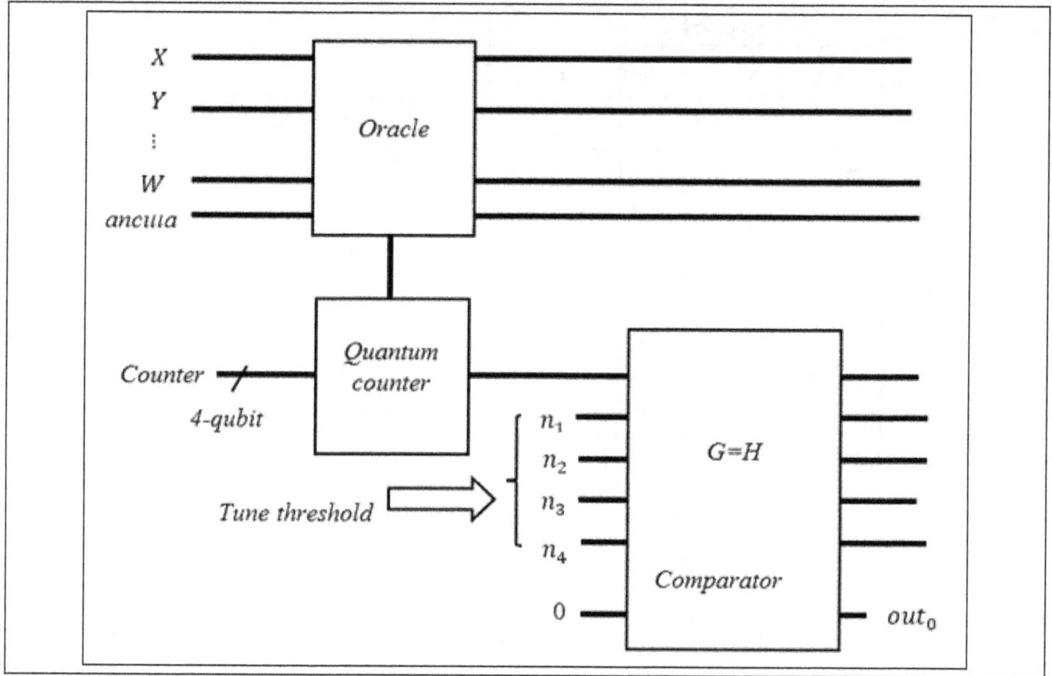

Figure 18: FSM Oracle Design with counter circuit and threshold with comparator

The solution with the minimum number of positive literals is $\{V, Q, R, Y, U\}$ which simplifies to $\{V, R, Y, U\}$ because group $Q = \{C, F\}$ is included in $U = \{C, E, F\}$, thus $Q \Rightarrow U$. Symbol V requires combining states A and C, symbol Y requires combining states B and F, symbol R combines states C and D and symbol U combines states C, E, F together. As the result, we obtain the minimized state machine from Figure 19a. We combine states from the respective states of Figure 15. Thus, combining in column 00 for row V we obtain symbols A and C and output 0. This way, combining states from groups V, Y, R and U the entire table from Figure 19a is created. Now for every subset of initial states A, B, C, D, E, F corresponding to each symbol from set of sets $\{V, Y, R, U\}$ we check to which set this subset belongs. For instance state C is included in sets V, R and U. Therefore symbol C in the table from Figure 19a is replaced with symbols V, R and U in the transition cells. This way, the non-deterministic machine from Figure 19b is created. Now select any state among V, R and U to create one of the deterministic machines described by the non-deterministic machine. Choose every row U in column 11 in order to improve the logic realization of the machine. Similarly, for the purpose of good encoding (encoding not explained here), select state R in column 00 to have two transitions to V and two transitions to R in this column. One final finite state machine minimized in this way is shown in Figure 19c.

a)

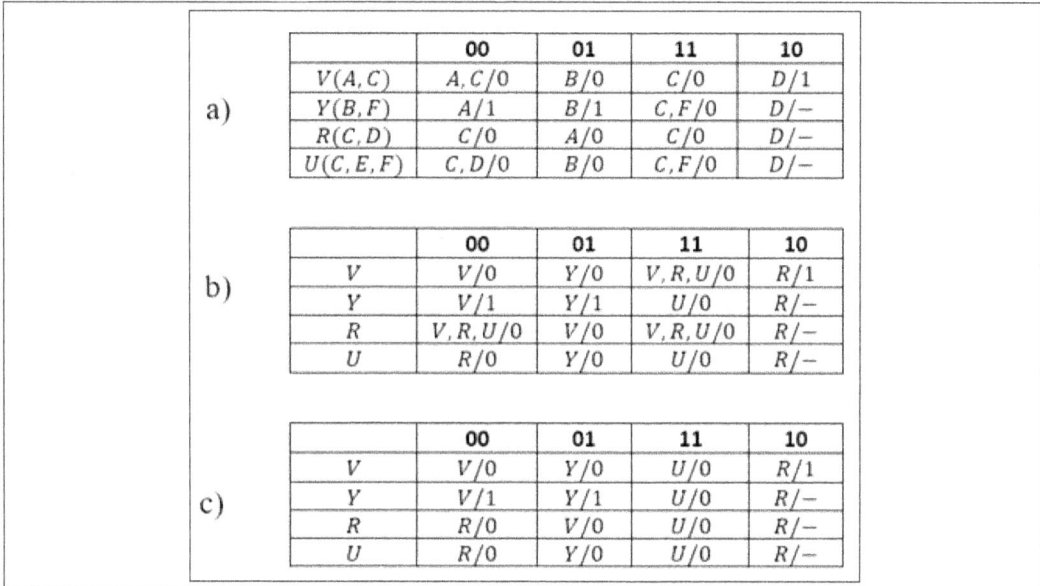

	00	01	11	10
V(A,C)	A,C/0	B/0	C/0	D/1
Y(B,F)	A/1	B/1	C,F/0	D/–
R(C,D)	C/0	A/0	C/0	D/–
U(C,E,F)	C,D/0	B/0	C,F/0	D/–

b)

	00	01	11	10
V	V/0	Y/0	V,R,U/0	R/1
Y	V/1	Y/1	U/0	R/–
R	V,R,U/0	V/0	V,R,U/0	R/–
U	R/0	Y/0	U/0	R/–

c)

	00	01	11	10
V	V/0	Y/0	U/0	R/1
Y	V/1	Y/1	U/0	R/–
R	R/0	V/0	U/0	R/–
U	R/0	Y/0	U/0	R/–

Figure 19: Steps to create an exactly minimized deterministic FSM using binate covering problem. (a) the table created directly from the solution to the covering-closure problem; (b) a non-deterministic automaton created from the table in (a); (c) one deterministic automaton created from the non-deterministic automaton in (b)

There are not yet benchmarks for quantum algorithms. However, there exist benchmarks for classical algorithms, such as those in [18, 22, 1, 9]. The current quantum computers are too small to run the classical benchmarks on them. One can, however, speculate on the speedup of future quantum and hybrid computers based on these classical benchmarks. Suppose a benchmark takes m terms and n variables, using our method, this benchmark would require n qubits for variables and $\lceil \log_2 m \rceil$ ancilla qubits for terms to represent the problem in a quantum algorithm design. In contrast, the traditional quantum oracle design would require n qubits for variables and $m+1$ ancilla qubits for terms [2]. Thus, when compared to the traditional Grover, our proposed design requires fewer qubits with a quadratic speedup of Grover's algorithm. As can be seen from Figure 20, for instance, if a given covering problem consists of 100,000 clauses, then our quantum oracle design requires only 18 ancilla qubits, while the traditional quantum oracle would require 100,000 ancilla qubits, which is the same as the clause number in the given problem. Assuming a complete search, the complexity of the classical algorithm would be $N = 2^n$. The complexity of our quantum algorithm would be $O(\sqrt{N})$. When the quantum computers have enough qubits, comparing practical benchmarks will be possible. Because IBM aims to build a quantum computer with 100,000 qubits in 10 years [10], we hope that in this time frame, our quantum algorithm for EDA problems will become practical.

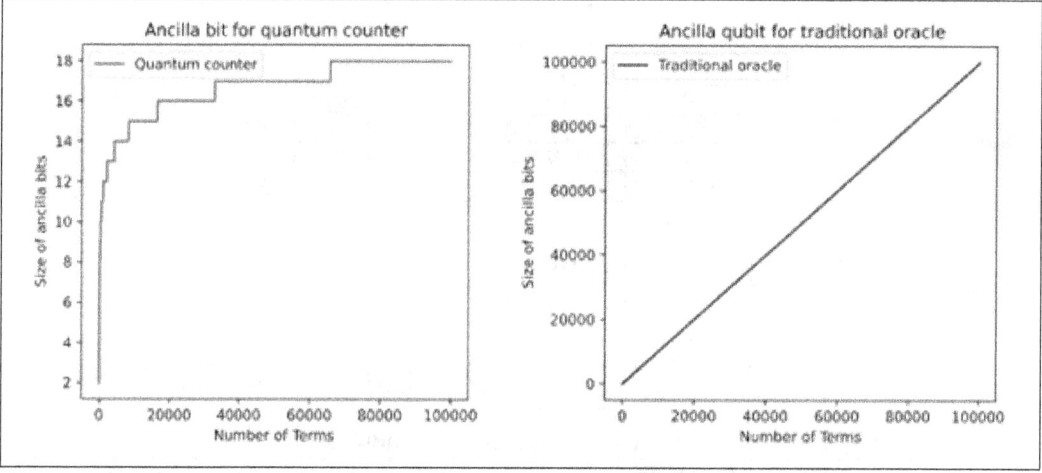

Figure 20: Comparison of the required numbers of ancilla qubits for our quantum oracle design (left) and the traditional quantum oracle design (right) [2]

6 Conclusion

We presented new quantum algorithms to solve with quadratic speedup several fundamental problems of classical logic circuits and finite state machine design. All these algorithms either use Grover, repeated Grover, or hybrid algorithms that use Grover's algorithm. Optimization problems are reduced to a repetition of constraint satisfaction problems solved by Grover' algorithm. Our approach is based on creating various oracles, which are, however, based on the same basic principle. We presented quantum oracle designs for various well-known EDA applications of the Unate and Binate covering problems. Innovative quantum algorithms for exact an incompletely specified FSM minimization have been presented here for the first time. Each of these problems is converted to a general quantum oracle. Our quantum oracle design uses an iterative quantum counter block used inside the oracles for the Grover-like algorithms. The concept of introducing this quantum counter can be applied to all these algorithms, allowing them to solve in a uniform way both SAT-like and MAX-SAT-like problems. Most importantly, this approach reduces the number of qubits logarithmically [3]. The introduction of the iterative quantum counter circuit replaces the ancilla qubits of the global large AND gate for traditional quantum oracle design for SAT-like problems. This is important because it reduces not only the number of required qubits but also avoids designing a quantum AND gate with many inputs, which is known to be a very complicated gate. We presented experimental results from the QISKIT simulator that showed the circuit works correctly.

Our future research will investigate using quantum counting based on combining Grover's algorithm with quantum phase estimation [29] to more specifically estimate the number of Grover's Loop repetitions for larger problems, which are difficult to calculate manually. Further improvement is to extend the quantum oracle design to add an arithmetic circuit for the weighted covering problems.

References

[1] The mcnc benchmark problems for vlsi floorplanning. `http://www.mcnc.org`. [Online].

[2] Abdirahman Alasow, Peter Jin, and Marek Perkowski. Quantum algorithm for variant maximum satisfiability. *Entropy*, 24(11), 2022.

[3] Abdirahman Alasow and Marek Perkowski. Quantum algorithm for maximum satisfiability. In *2022 IEEE 52nd International Symposium on Multiple-Valued Logic (ISMVL)*, pages 27–34. IEEE, 2022.

[4] Abdirahman Alasow and Marek Perkowski. Quantum algorithm for mining frequent patterns for association rule mining. *Journal of Quantum Information Science*, 13(1):1–23, 2023.

[5] Gadi Aleksandrowicz, Thomas Alexander, Panagiotis Barkoutsos, Luciano Bello, Yael Ben-Haim, David Bucher, F Jose Cabrera-Hernández, Jorge Carballo-Franquis, Adrian Chen, Chun-Fu Chen, et al. Qiskit: An open-source framework for quantum computing. *Accessed on: Mar*, 16, 2019.

[6] Carolina Allende, Efrain Buksman, and André Fonseca De Oliveira. Quantum circuit design using neural networks assisted by entanglement. In *2021 IEEE URUCON*, pages 316–319. IEEE, 2021.

[7] Julien Bramel and David Simchi-Levi. On the effectiveness of set covering formulations for the vehicle routing problem with time windows. *Operations Research*, 45(2):295–301, 1997.

[8] Melvin A Breuer. *Design Automation of Digital Systems: Vol. 1.: Theory and Techniques.* Prentice-Hall, 1972.

[9] Franc Brglez. A neutral netlist of 10 combinational benchmark circuits and a target translator in fortran. In *Proc. Intl. Symp. Circuits and Systems, 1985*, 1985.

[10] Michael Brooks. Ibm wants to build a 100,000-qubit quantum computer. `https://www.technologyreview.com/2023/05/25/1073606/ibm-wants-to-build-a-100000-qubit-quantum-computer`. Accessed: 2023-06-15.

[11] Hongxiang Fan, Ce Guo, and Wayne Luk. Optimizing quantum circuit placement via machine learning. In *Proceedings of the 59th ACM/IEEE Design Automation Conference*, pages 19–24, 2022.

[12] Gary D Hachtel and Fabio Somenzi. *Logic synthesis and verification algorithms.* Springer Science & Business Media, 2005.

[13] Chi-Chuan Hwang, Chu-Yuan Tseng, and Cheng-Fang Su. Quantum circuit design for computer-assisted shor's algorithm. 2022.

[14] Tommy R Jensen and Bjarne Toft. *Graph coloring problems.* John Wiley & Sons, 2011.

[15] Jeong and Somenzi. A new algorithm for the binate covering problem and its application to the minimization of boolean relations. In *1992 IEEE/ACM International Conference on Computer-Aided Design*, pages 417–420. IEEE, 1992.

[16] Ankit Kagliwal and Shankar Balachandran. Set-cover heuristics for two-level logic minimization. In *2012 25th International Conference on VLSI Design*, pages 197–202. IEEE, 2012.

[17] Timothy Kam, Tiziano Villa, Robert Brayton, and Alberto Sangiovanni-Vincentelli. A fully implicit algorithm for exact state minimization. In *Proceedings of the 31st annual Design Automation Conference*, pages 684–690, 1994.

[18] Timothy Kam, Tiziano Villa, Robert K Brayton, and Alberto L Sangiovanni-Vincentelli. *Synthesis of finite state machines: functional optimization.* Springer Science & Business Media, 2013.

[19] Sumeet Khatri, Ryan LaRose, Alexander Poremba, Lukasz Cincio, Andrew T Sornborger, and Patrick J Coles. Quantum-assisted quantum compiling. *Quantum*, 3:140, 2019.

[20] Zvi Kohavi and Niraj K Jha. *Switching and finite automata theory.* Cambridge University Press, 2009.

[21] Xiao Yu Li. *Optimization algorithms for the minimum-cost satisfiability problem.* North Carolina State University, 2004.

[22] Xiao Yu Li, Matthias F Stallmann, and Franc Brglez. Effective bounding techniques for solving unate and binate covering problems. In *Proceedings of the 42nd annual Design Automation Conference*, pages 385–390, 2005.

[23] Yiwei Li, Edison Tsai, Marek Perkowski, and Xiaoyu Song. Grover-based ashenhurst-curtis decomposition using quantum language quipper. *Quantum Information & Computation*, 19(1-2):35–66, 2019.

[24] Stan Liao, Srinivas Devadas, Kurt Keutzer, and Steve Tjiang. Instruction selection using binate covering for code size optimization. In *Proceedings of IEEE International Conference on Computer Aided Design (ICCAD)*, pages 393–399. IEEE, 1995.

[25] Stan Liao, Kurt Keutzer, STEVEN Tjiang, and Srinivas Devadas. A new viewpoint on code generation for directed acyclic graphs. *ACM Transactions on Design Automation of Electronic Systems (TODAES)*, 3(1):51–75, 1998.

[26] Edward J McCluskey. Minimization of boolean functions. *The Bell System Technical Journal*, 35(6):1417–1444, 1956.

[27] Marta Mesquita and Ana Paias. Set partitioning/covering-based approaches for the integrated vehicle and crew scheduling problem. *Computers & Operations Research*, 35(5):1562–1575, 2008.

[28] Giovanni De Micheli. *Synthesis and optimization of digital circuits*. McGraw-Hill Higher Education, 1994.

[29] Michael A Nielsen and Isaac L Chuang. *Quantum computation and quantum information*. Cambridge university press, 2010.

[30] Eric Paul, Bernd Steinbach, and Marek Perkowski. Application of cuda in the boolean domain for the unate covering problem. 2010.

[31] Marek Perkowski. Inverse problems, constraint satisfaction, reversible logic, invertible logic and grover quantum oracles for practical problems. In *Reversible Computation: 12th International Conference, RC 2020, Oslo, Norway, July 9-10, 2020, Proceedings 12*, pages 3–32. Springer, 2020.

[32] Marek A Perkowski, Jiuling Liu, and James E Brown. Rapid software prototyping: Cad design of digital cad algorithms. In *Progress in computer-aided VLSI design*, pages 353–401. 1990.

[33] Stanley R Petrick. A direct determination of the irredundant forms of a boolean function from the set of prime implicants. *Air Force Cambridge Res. Center Tech. Report*, pages 56–110, 1956.

[34] Richard L Rudell. *Multiple-valued logic minimization for PLA synthesis*. Electronics Research Laboratory, College of Engineering, University of ..., 1986.

[35] Richard L Rudell. *Logic synthesis for VLSI design*. University of California, Berkeley, 1989.

[36] Tsutomu Sasao. *Logic synthesis and optimization*, volume 2. Springer, 1993.

[37] Michal Servit and Jan Zamazal. Heuristic approach to binate covering problem. In *Proceedings The European Conference on Design Automation*, pages 123–124. IEEE Computer Society, 1992.

[38] Bernd Steinbach and Christian Posthoff. Sources and obstacles for parallelization-a comprehensive exploration of the unate covering problem using both cpu and gpu. *GPU Computing with Applications in Digital Logic*, 63, 2012.

[39] Bernd Steinbach and Christian Posthoff. Fast calculation of exact minimal unate coverings on both the cpu and the gpu. In *Computer Aided Systems Theory-EUROCAST 2013: 14th International Conference, Las Palmas de Gran Canaria, Spain, February*

10-15, 2013. Revised Selected Papers, Part II 14, pages 234–241. Springer, 2013.

[40] Bernd Steinbach and Matthias Werner. Alternative approaches for fast boolean calculations using the gpu. In *Computational Intelligence and Efficiency in Engineering Systems*, pages 17–31. Springer, 2015.

[41] Tiziano Villa, Timothy Kam, Robert K Brayton, and Alberto L Sangiovanni-Vincenteili. Explicit and implicit algorithms for binate covering problems. *IEEE Transactions on computer-Aided Design of integrated Circuits and Systems*, 16(7):677–691, 1997.

[42] Wikipedia. Grover's algorithm — Wikipedia, the free encyclopedia. `http://en.wikipedia.org/w/index.php?title=Grover's%20algorithm&oldid=1169154668`. accessed: 2023-06-27.

Received

EMBEDDING FIRST-ORDER CLASSICAL LOGIC INTO GUREVICH'S EXTENDED FIRST-ORDER INTUITIONISTIC LOGIC: THE ROLE OF STRONG NEGATION

NORIHIRO KAMIDE
Nagoya City University, School of Data Science,
Yamanohata 1, Mizuho-cho, Mizuho-ku, Nagoya, Aichi, 467-8501, Japan
drnkamide08@kpd.biglobe.ne.jp

Abstract

In this study, a theorem for embedding first-order classical logic into Gurevich's extended first-order intuitionistic logic with strong negation is investigated in terms of the Gödel–Gentzen negative translation. First, an alternative cut-free Gentzen-style sequent calculus ELK for first-order classical logic is introduced to extend Gentzen's sequent calculus LK for first-order classical logic. Second, a theorem for embedding ELK into a Gentzen-style sequent calculus ELJ for Gurevich's extended first-order intuitionistic logic with strong negation is proved using an extended Gödel–Gentzen negative translation. Finally, a theorem for embedding ELK into Gentzen's sequent calculus LJ for first-order intuitionistic logic is obtained using a slightly modified version of the extended Gödel–Gentzen negative translation.

1 Introduction

In this study, we investigate a theorem for embedding *first-order classical logic* (CL) into *Gurevich's extended first-order intuitionistic logic with strong negation* (GL) [12]. First, to investigate the embedding theorem, we introduce an alternative cut-free Gentzen-style sequent calculus ELK for CL by extending Gentzen's sequent calculus LK [10] for CL. Second, we introduce an extended version of the well-known *Gödel–Gentzen negative translation* [11, 9] from CL to *first-order intuitionistic logic*

We would like to thank the referees for their valuable comments. This research was supported by JSPS KAKENHI Grant Number 23K10990.

(IL). Third, we prove a theorem for embedding ELK into a Gentzen-style sequent calculus ELJ [15] for GL by using the extended Gödel–Gentzen negative translation (i.e., we obtain a theorem for embedding CL into GL). Finally, we obtain a theorem for embedding ELK into Gentzen's sequent calculus LJ for IL by using a slightly modified version of the extended Gödel–Gentzen negative translation (i.e., we obtain an alternative new theorem for embedding CL into IL). Based on these results, we can clarify and understand the role of *strong negation* in CL and GL, where strong negation was introduced in [19] and traditionally used in GL and related constructive logics [1, 19, 21, 28, 26, 30]. More specifically, we can clarify and understand that strong negation is an essential and natural component of CL for representing falsification-aware reasoning and plays a crucial role in smoothly proving some theorems for embedding CL into GL and IL (and some related theorems for CL).

The key technique for smoothly proving the theorem for embedding CL into GL is to construct an alternative cut-free Gentzen-style sequent calculus ELK for CL. We first explain the concept of constructing ELK. We construct ELK as a natural classical extension or a version of ELJ, which was introduced in [15] for GL (although it was referred to as GI in [15]). This implies that the language of CL is extended by adding a classical strong negation connective, which is regarded as the classical counterpart of the (intuitionistic) strong negation connective equipped with GL. Thus, this extended language includes two types of negation connectives: $\sim$ (classical strong negation) and $\neg$ (classical negation). However, we can show that these negations are essentially equivalent in ELK (i.e., we can prove that ELK $\vdash$ ${\sim}\alpha \Rightarrow \neg\alpha$ and ELK $\vdash \neg\alpha \Rightarrow {\sim}\alpha$). This means that $\sim$ (classical strong negation) is equivalent to $\neg$ (classical negation) in the context of CL. However, as is well known, $\sim$ (strong negation) is not equivalent to $\neg$ (intuitionistic negation) in the context of IL (i.e., we can prove that ELJ $\vdash$ ${\sim}\alpha \Rightarrow \neg\alpha$, but we cannot prove that ELJ $\vdash \neg\alpha \Rightarrow {\sim}\alpha$ in general; hence $\sim$ is "stronger" than $\neg$). Thus, $\sim$ (classical strong negation) is regarded as a redundant and auxiliary negation connective in ELK, but this connective plays a crucial role in proving the theorem for embedding ELK into ELJ.

ELK is also constructed as an extension and combination of both Gentzen's LK and the falsification-aware normal Gentzen-style sequent calculus LKF introduced in [16]. LKF is obtained from ELK by deleting the logical inference rules concerning $\neg$, and LK is obtained from ELK by deleting the initial sequents and logical inference rules concerning $\sim$. On the one hand, the cut-elimination and completeness theorems for LK are well known [10, 25]. On the other hand, the cut-elimination and completeness theorems for LKF were proved in [16] using Schütte's method. The theorem-equivalence between LK and LKF was also proved in [16]. Thus, LK, LKF, and ELK are all Gentzen-style sequent calculi for CL. In this study, we prove a weak theorem for syntactically embedding ELK into LK and use this theorem to prove the cut-elimination theorem for ELK. Using this cut-elimination theorem, we can also obtain the cut-elimination theorem for LKF because LKF is a subsystem of ELK. This fact provides a solution to the unsolved problem discussed in [16] for obtaining an embedding-based syntactical proof of the cut-elimination theorem for LKF. Thus, the proposed embedding-based simple syntactical proof of the cut-elimination theorems for ELK and LKF is a novel contribution of this study. More specifically, we can understand that the use of $\sim$ (classical strong negation) plays a crucial role in smoothly obtaining the embedding-based proof of the cut-elimination theorems for ELK and LKF.

Next, we provide the background of GL. GL was originally introduced by Gurevich in [12], wherein a Hilbert-style axiomatic system, a cut-free Gentzen-style sequent calculus, and Kripke and three-valued semantics were introduced for GL. In addition, the completeness theorems with respect to the Kripke and three-valued semantics were proved for GL. As previously mentioned, ELJ was introduced in [15] as an alternative Gentzen-style sequent calculus for GL. ELJ and the original system $\overline{G}$ by Gurevich for GL were shown to be theorem-equivalent. The Hilbert-style axiomatic system introduced by Gurevich was obtained from that for IL by adding the following axiom schemes, where $\sim$ and $\neg$ are the strong negation and intuitionistic negation connectives, respectively:

1. $\sim\sim\alpha \leftrightarrow \alpha$,
2. $\sim\neg\alpha \leftrightarrow \alpha$,
3. $\sim\alpha\rightarrow\neg\alpha$,
4. $\sim(\alpha\wedge\beta) \leftrightarrow \sim\alpha\vee\sim\beta$,
5. $\sim(\alpha\vee\beta) \leftrightarrow \sim\alpha\wedge\sim\beta$,
6. $\sim(\alpha\rightarrow\beta) \leftrightarrow \alpha\wedge\sim\beta$,
7. $\sim\forall x\alpha \leftrightarrow \exists x\sim\alpha$,
8. $\sim\exists x\alpha \leftrightarrow \forall x\sim\alpha$.

This axiomatic system is theorem-equivalent to the axiomatic system obtained from this system by replacing the law $\sim\alpha\rightarrow\neg\alpha$ of strong negation with the law $(\alpha\wedge\sim\alpha)\rightarrow\beta$ of explosion. For more information on this, see Remark 2.20 in Section 2 of this paper.

Next, we explain the aims of this study. One aim is to clarify the relationship between GL and CL by extending the Gödel–Gentzen negative translation by means of strong negation. Through this clarification, we can understand the significance of GL and its related subsystem N3, which are known to be typical logics with strong negation. In addition, through the theorem for embedding ELK (i.e., CL) into ELJ (i.e., GL), GL can be considered as an alternative first-order intuitionistic logic. Another aim is to construct an alternative Gödel–Gentzen negative translation from CL to IL by adding strong negation. Through this construction, we can understand the role and essence of strong negation behind CL and IL (i.e., we can obtain another new perspective on CL and IL by means of strong negation).

The remainder of this paper is organized as follows. In Section 2, we introduce ELK with an extended auxiliary language that includes $\sim$ (classical strong negation) to extend LK. In addition, some theorems for embedding ELK into LK are proved and the cut-elimination theorem for ELK is then proved using one of these embedding theorems. Finally, in this same section, we introduce ELJ and provide some remarks on ELJ. In Section 3, we prove a theorem for embedding ELK into ELJ using an extended Gödel–Gentzen negative translation, and we derive a theorem for embedding ELK into LJ using a slightly modified version of the extended Gödel–Gentzen negative translation. In Section 4, we conclude this paper and address some remarks on related works.

2 Sequent calculi

2.1 Alternative sequent calculus for classical logic

Formulas of CL with an extended language including classical strong negation are constructed using countably many predicate symbols $p, q, ...$, countably many individual variables $x, y, ...$, countably many individual constants $a, b, ...$, countably many function symbols $s, t, ...$, and the logical connectives $\wedge$ (conjunction), $\vee$ (disjunction), $\rightarrow$ (implication), $\neg$ (classical negation), $\sim$ (classical strong negation), $\forall$ (universal quantifier), and $\exists$ (existential quantifier). *Terms* are constructed from individual variables, individual constants, and function symbols. We use small letters $p, q, ...$ to denote not only predicate symbols but also atomic formulas, small letters $s, t, ...$ to denote not only function symbols but also terms, Greek small letters $\alpha, \beta, ...$ to denote formulas, and Greek capital letters $\Gamma, \Delta, ...$ to represent finite (possibly empty) sets of formulas. We use an expression $\alpha[t/x]$ to represent the formula that is obtained from the formula α by replacing all free occurrences of the individual variable x in α by the term t, but avoiding a clash of variables by a suitable renaming of bound variables. We use the symbol $\equiv$ to denote the equality of symbols. We call an expression of the form $\Gamma \Rightarrow \Delta$ *sequent*. We use an expression $L \vdash \Gamma \Rightarrow \Delta$ to represent the fact that the sequent $\Gamma \Rightarrow \Delta$ is provable in a Gentzen-style sequent calculus L where L in this expression will occasionally be omitted. We use an expression $\alpha \Leftrightarrow \beta$ to represent the abbreviation of the sequents $\alpha \Rightarrow \beta$ and $\beta \Rightarrow \alpha$. We say that "two Gentzen-style sequent calculi L_1 and L_2 are *theorem-equivalent*" if $\{S \mid L_1 \vdash S\} = \{S \mid L_2 \vdash S\}$. We say that "a rule R of inference is *admissible* in a Gentzen-style sequent calculus L" if the following condition is satisfied: For any instance

$$\frac{S_1 \cdots S_n}{S}$$

of R, if $L \vdash S_i$ for all i, then $L \vdash S$. Furthermore, we say that "R is *derivable* in L" if there is a derivation from $S_1, \cdots, S_n$ to S in L.

We now introduce a Gentzen-style sequent calculus ELK for CL with the extended language including $\sim$.

Definition 2.1 (ELK). *In the following definition, we use the symbol p to represent an arbitrary atomic formula, the symbol t to represent an arbitrary term, and the symbol z to represent a special individual variable (called* eigenvariable*) that satisfies the following condition: z does not occur as a free individual variable in the lower sequent of the rule.*

The initial sequents of ELK *are of the form:*

$$p \Rightarrow p \qquad {\sim}p \Rightarrow {\sim}p \qquad p, {\sim}p \Rightarrow \qquad \Rightarrow p, {\sim}p.$$

The structural inference rules of ELK *are of the form:*

$$\frac{\Gamma \Rightarrow \Delta, \alpha \quad \alpha, \Sigma \Rightarrow \Pi}{\Gamma, \Sigma \Rightarrow \Delta, \Pi} \text{(cut)} \quad \frac{\Gamma \Rightarrow \Delta}{\alpha, \Gamma \Rightarrow \Delta} \text{(we-left)} \quad \frac{\Gamma \Rightarrow \Delta}{\Gamma \Rightarrow \Delta, \alpha} \text{(we-right)}.$$

The normal logical inference rules of ELK *are of the form:*

$$\frac{\alpha, \beta, \Gamma \Rightarrow \Delta}{\alpha \wedge \beta, \Gamma \Rightarrow \Delta} (\wedge\text{left}) \quad \frac{\Gamma \Rightarrow \Delta, \alpha \quad \Gamma \Rightarrow \Delta, \beta}{\Gamma \Rightarrow \Delta, \alpha \wedge \beta} (\wedge\text{right})$$

$$\frac{\alpha, \Gamma \Rightarrow \Delta \quad \beta, \Gamma \Rightarrow \Delta}{\alpha \vee \beta, \Gamma \Rightarrow \Delta} (\vee\text{left}) \quad \frac{\Gamma \Rightarrow \Delta, \alpha, \beta}{\Gamma \Rightarrow \Delta, \alpha \vee \beta} (\vee\text{right})$$

$$\frac{\Gamma \Rightarrow \Delta, \alpha \quad \beta, \Sigma \Rightarrow \Pi}{\alpha \to \beta, \Gamma, \Sigma \Rightarrow \Delta, \Pi} (\to\text{left}) \quad \frac{\alpha, \Gamma \Rightarrow \Delta, \beta}{\Gamma \Rightarrow \Delta, \alpha \to \beta} (\to\text{right})$$

$$\frac{\Gamma \Rightarrow \Delta, \alpha}{\neg\alpha, \Gamma \Rightarrow \Delta} (\neg\text{left}) \quad \frac{\alpha, \Gamma \Rightarrow \Delta}{\Gamma \Rightarrow \Delta, \neg\alpha} (\neg\text{right})$$

$$\frac{\alpha[t/x], \Gamma \Rightarrow \Delta}{\forall x\alpha, \Gamma \Rightarrow \Delta} \text{ } (\forall\text{left}) \quad \frac{\Gamma \Rightarrow \Delta, \alpha[z/x]}{\Gamma \Rightarrow \Delta, \forall x\alpha} \text{ } (\forall\text{right})$$

$$\frac{\alpha[z/x], \Gamma \Rightarrow \Delta}{\exists x\alpha, \Gamma \Rightarrow \Delta} \text{ } (\exists\text{left}) \quad \frac{\Gamma \Rightarrow \Delta, \alpha[t/x]}{\Gamma \Rightarrow \Delta, \exists x\alpha} \text{ } (\exists\text{right}).$$

The strongly-negated logical inference rules of ELK *are of the form:*

$$\frac{\alpha, \Gamma \Rightarrow \Delta}{{\sim}{\sim}\alpha, \Gamma \Rightarrow \Delta} \text{ } ({\sim}{\sim}\text{left}) \quad \frac{\Gamma \Rightarrow \Delta, \alpha}{\Gamma \Rightarrow \Delta, {\sim}{\sim}\alpha} \text{ } ({\sim}{\sim}\text{right})$$

$$\frac{{\sim}\alpha, \Gamma \Rightarrow \Delta \quad {\sim}\beta, \Gamma \Rightarrow \Delta}{{\sim}(\alpha \wedge \beta), \Gamma \Rightarrow \Delta} \text{ } ({\sim}\wedge\text{left}) \quad \frac{\Gamma \Rightarrow \Delta, {\sim}\alpha, {\sim}\beta}{\Gamma \Rightarrow \Delta, {\sim}(\alpha \wedge \beta)} \text{ } ({\sim}\wedge\text{right})$$

$$\frac{{\sim}\alpha, {\sim}\beta, \Gamma \Rightarrow \Delta}{{\sim}(\alpha \vee \beta), \Gamma \Rightarrow \Delta} \text{ } ({\sim}\vee\text{left}) \quad \frac{\Gamma \Rightarrow \Delta, {\sim}\alpha \quad \Gamma \Rightarrow \Delta, {\sim}\beta}{\Gamma \Rightarrow \Delta, {\sim}(\alpha \vee \beta)} \text{ } ({\sim}\vee\text{right})$$

$$\frac{\alpha, {\sim}\beta, \Gamma \Rightarrow \Delta}{{\sim}(\alpha \rightarrow \beta), \Gamma \Rightarrow \Delta} \text{ } ({\sim}{\rightarrow}\text{left}) \quad \frac{\Gamma \Rightarrow \Delta, \alpha \quad \Gamma \Rightarrow \Delta, {\sim}\beta}{\Gamma \Rightarrow \Delta, {\sim}(\alpha \rightarrow \beta)} \text{ } ({\sim}{\rightarrow}\text{right})$$

$$\frac{\alpha, \Gamma \Rightarrow \Delta}{{\sim}\neg\alpha, \Gamma \Rightarrow \Delta} \text{ } ({\sim}\neg\text{left}) \quad \frac{\Gamma \Rightarrow \Delta, \alpha}{\Gamma \Rightarrow \Delta, {\sim}\neg\alpha} \text{ } ({\sim}\neg\text{right})$$

$$\frac{{\sim}\alpha[z/x], \Gamma \Rightarrow \Delta}{{\sim}\forall x\alpha, \Gamma \Rightarrow \Delta} \text{ } ({\sim}\forall\text{left}) \quad \frac{\Gamma \Rightarrow \Delta, {\sim}\alpha[t/x]}{\Gamma \Rightarrow \Delta, {\sim}\forall x\alpha} \text{ } ({\sim}\forall\text{right})$$

$$\frac{{\sim}\alpha[t/x], \Gamma \Rightarrow \Delta}{{\sim}\exists x\alpha, \Gamma \Rightarrow \Delta} \text{ } ({\sim}\exists\text{left}) \quad \frac{\Gamma \Rightarrow \Delta, {\sim}\alpha[z/x]}{\Gamma \Rightarrow \Delta, {\sim}\exists x\alpha} \text{ } ({\sim}\exists\text{right}).$$

Remark 2.2. *For the sake of simplicity,* ELK *adopts the explicit weakening rules* (we-left) *and* (we-right), *but do not adopt the contraction and exchange rules. Concerned with this setting,* ELK *adopts the set-based sequents* $\Gamma \Rightarrow \Delta$ *where* Γ *and* Δ *are (possibly empty) sets of formulas. Similar setting will be used for the sequent calculus* ELJ *for Gurevich logic.*

Proposition 2.3. *Sequents of the form* $\alpha \Rightarrow \alpha$ *for any formula* α *are provable in cut-free* ELK.

Proof. By induction on α. ☐

Proposition 2.4. *Sequents of the form* $\alpha, {\sim}\alpha \Rightarrow$ *and* $\Rightarrow \alpha, {\sim}\alpha$ *for any formula* α *are provable in cut-free* ELK.

Proof. By induction on α. We distinguish the cases according to the form of α and show only the following cases.

1. Case $\alpha \equiv \beta{\to}\gamma$: We obtain the required fact:

$$\dfrac{\dfrac{\begin{matrix}\vdots\ \textit{Prop. 2.3}\\ \beta \Rightarrow \beta\end{matrix} \qquad \begin{matrix}\vdots\ \textit{Ind. hyp.}\\ \gamma, {\sim}\gamma \Rightarrow\end{matrix}}{\beta{\to}\gamma, \beta, {\sim}\gamma \Rightarrow}\,(\to\text{left})}{\beta{\to}\gamma, {\sim}(\beta{\to}\gamma) \Rightarrow}\,({\sim}{\to}\text{left})$$

$$\dfrac{\dfrac{\dfrac{\begin{matrix}\vdots\ \textit{Prop. 2.3}\\ \beta \Rightarrow \beta\end{matrix}}{\beta \Rightarrow \gamma, \beta}\,(\text{we-right})}{\Rightarrow \beta{\to}\gamma, \beta}\,(\to\text{left}) \qquad \dfrac{\dfrac{\begin{matrix}\vdots\ \textit{Ind. hyp.}\\ \Rightarrow \gamma, {\sim}\gamma\end{matrix}}{\beta \Rightarrow \gamma, {\sim}\gamma}\,(\text{we-left})}{\Rightarrow \beta{\to}\gamma, {\sim}\gamma}\,(\to\text{right})}{\Rightarrow \beta{\to}\gamma, {\sim}(\beta{\to}\gamma)}\,({\sim}{\to}\text{right}).$$

2. Case $\alpha \equiv \neg\beta$: We obtain the required fact:

$$\dfrac{\dfrac{\begin{matrix}\vdots\ \textit{Prop. 2.3}\\ \beta \Rightarrow \beta\end{matrix}}{\neg\beta, \beta \Rightarrow}\,(\neg\text{left})}{\neg\beta, {\sim}\neg\beta \Rightarrow}\,({\sim}\neg\text{left}) \qquad \dfrac{\dfrac{\begin{matrix}\vdots\ \textit{Prop. 2.3}\\ \beta \Rightarrow \beta\end{matrix}}{\Rightarrow \neg\beta, \beta}\,(\neg\text{right})}{\Rightarrow \neg\beta, {\sim}\neg\beta}\,({\sim}\neg\text{right})$$

□

Proposition 2.5. *The following sequents are provable in cut-free* ELK*: For any formulas α and β,*

1. ${\sim}{\sim}\alpha \Leftrightarrow \alpha$,
2. ${\sim}(\alpha\wedge\beta) \Leftrightarrow {\sim}\alpha\vee{\sim}\beta$,
3. ${\sim}(\alpha\vee\beta) \Leftrightarrow {\sim}\alpha\wedge{\sim}\beta$,
4. ${\sim}(\alpha{\to}\beta) \Leftrightarrow \alpha\wedge{\sim}\beta$,
5. ${\sim}\neg\alpha \Leftrightarrow \alpha$,
6. ${\sim}\forall x\alpha \Leftrightarrow \exists x{\sim}\alpha$,
7. ${\sim}\exists x\alpha \Leftrightarrow \forall x{\sim}\alpha$,
8. $\alpha\wedge{\sim}\alpha \Rightarrow \beta$,
9. $\Rightarrow \alpha\vee{\sim}\alpha$,

10. $\sim\alpha \Leftrightarrow \neg\alpha$.

Proof. By using Propositions 2.3 and 2.4. To prove the cases (8) and (9), Proposition 2.4 is required. We show only the case (10) as follows.

$$\dfrac{\begin{array}{c}\vdots\ Prop.\ 2.4\\ \alpha, \sim\alpha \Rightarrow\end{array}}{\sim\alpha \Rightarrow \neg\alpha}\ (\neg\text{right}) \qquad \dfrac{\begin{array}{c}\vdots\ Prop.\ 2.4\\ \Rightarrow \alpha, \sim\alpha\end{array}}{\neg\alpha \Rightarrow \sim\alpha}\ (\neg\text{left}).$$

□

Next, we introduce a Gentzen-style sequent calculus LK for CL. The language of LK is obtained from that of ELK by deleting $\sim$.

Definition 2.6 (LK). *A Gentzen-style sequent calculus* LK *for* CL *is defined as the* $\sim$*-free part of* ELK *(i.e., it is obtained from* ELK *by deleting the strongly-negated initial sequents and strongly-negated logical inference rules).*

Remark 2.7. *We make the following remarks.*

1. *Proposition 2.5 (10) means that* $\sim$ *and* $\neg$ *are equivalent. Namely, strong negation in the context of* CL *is equivalent to classical negation. Moreover, by Proposition 2.5 (10), we can understand that* ELK *and* LK *are essentially equivalent. Namely,* ELK *is regarded as an alternative Gentzen-style sequent calculus for* CL *with an extended auxiliary language including* $\sim$.

2. *On the one hand,* $\sim$ *and* $\neg$ *are equivalent in the context of* CL*. On the other hand,* $\sim$ *and* $\neg$ *are not equivalent in the context of* IL *(i.e., strong negation in the context of* IL *is not equivalent to intuitionistic negation). Thus, the classical logic counterpart* $\sim$ *of strong negation is required for showing a theorem for embedding* ELK *into a Gentzen-style sequent calculus* ELJ *for* GL [12]*. Actually,* $\sim$ *has a crucial role for smoothly proving the theorem for embedding* ELK *into* ELJ.

3. *The system* ELJ *was introduced in* [15] *and was also referred to as* GI *in* [15]*.* ELJ *will be precisely defined as an intuitionistic version (or subsystem) of* ELK *that is characterized by the intuiitionistic sequents of the form* $\Gamma \Rightarrow \gamma$ *where* γ *is a single formula or the empty set. The subsystem that is obtained from* ELJ *by deleting all the logical inference rules concerning* $\neg$ *is a Gentzen-style sequent calculus for Nelson's three-valued constructive logic (*N3*)* [1, 19].

4. *The subsystem that is obtained from* ELK *by deleting the logical inference rules concerning* $\neg$ *was referred to as* LKF *in* [16] *and was defined as a "falsification-aware" Gentzen-style sequent calculus for* CL. *The cut-elimination and completeness theorems for* LKF *were proved in* [16] *using Schütte's method. The theorem-equivalence between* LKF *and* LK *was also proved in* [16]. *Thus, the systems* LKF, ELK, *and* LK *are all Gentzen-style sequent calculi for* CL.

5. *The subsystem that is obtained from* ELK *by deleting the logical inference rules concerning* $\neg$ *and the initial sequents* $p, {\sim}p \Rightarrow$ *and* $\Rightarrow p, {\sim}p$ *(i.e., it is obtained from* LKF *by deleting* $p, {\sim}p \Rightarrow$ *and* $\Rightarrow p, {\sim}p$*) is regarded as a Gentzen-style sequent calculus for the* $\{\wedge, \vee, \rightarrow, {\sim}\}$*-fragment of Arieli–Avron logics of logical bilattices* [2, 3]. *The intuitionistic version of this subsystem is a Gentzen-style sequent calculus for Nelson's four-valued constructive logic (*N4*)* [1, 19]. *For more information on Gentzen-style sequent calculi for* N4, *see* [17, 18].

6. *The name "*LK*" used in this paper is from Gentzen's original sequent calculus* LK *for* CL [10]. *In this paper, the name* LK *is used for a small modification of the original* LK. *The same proposition as Proposition 2.3 holds for* LK. *The cut-elimination theorem for* LK *holds. For more information on Gentzen's* LK, *consult e.g.,* [10, 25].

Next, we show the cut-elimination theorem for ELK by using a theorem for embedding ELK into LK. Prior to show the embedding theorem, we introduce a translation form the formulas of ELK to those of LK.

Definition 2.8. *Let* Φ *be a set of atomic formulas (or predicate symbols). The language (or the set of formulas)* $\mathcal{L}_{\mathrm{ELK}}$ *of* ELK *is defined using terms,* Φ, $\wedge$, $\vee$, $\rightarrow$, $\neg$, ${\sim}$, $\forall$, *and* $\exists$. *The language (or the set of formulas)* $\mathcal{L}_{\mathrm{LK}}$ *of* LK *is obtained from* $\mathcal{L}_{\mathrm{ELK}}$ *by deleting* ${\sim}$.

A mapping f *from* $\mathcal{L}_{\mathrm{ELK}}$ *to* $\mathcal{L}_{\mathrm{LK}}$ *is defined inductively by:*

1. *For any* $p \in \Phi$, $f(p) := p$ *and* $f({\sim}p) := \neg p$,

2. $f(\alpha \sharp \beta) := f(\alpha) \sharp f(\beta)$ *where* $\sharp \in \{\wedge, \vee, \rightarrow\}$,

3. $f(\sharp \alpha) := \sharp f(\alpha)$ *where* $\sharp \in \{\neg, \forall x, \exists x\}$,

4. $f(\sharp \alpha) := f(\alpha)$ *where* $\sharp \in \{{\sim}{\sim}, {\sim}\neg\}$,

5. $f({\sim}(\alpha \wedge \beta)) := f({\sim}\alpha) \vee f({\sim}\beta)$,

6. $f({\sim}(\alpha \vee \beta)) := f({\sim}\alpha) \wedge f({\sim}\beta)$,

7. $f({\sim}(\alpha{\rightarrow}\beta)) := f(\alpha){\wedge}f({\sim}\beta)$,

8. $f({\sim}\forall x\alpha) := \exists x f({\sim}\alpha)$,

9. $f({\sim}\exists x\alpha) := \forall x f({\sim}\alpha)$.

An expression $f(\Gamma)$ denotes the result of replacing every occurrence of a formula α in Γ by an occurrence of $f(\alpha)$. Analogous notation is used for another mapping discussed later.

Remark 2.9. *We make the following remarks.*

1. *The translation defined in Definition 2.8 is independent of terms (i.e., terms are not changed by this translation).*

2. *The translation function defined in Definition 2.8 is a modified extension of the translation function from* LKF *to* LK, *which was introduced in* [16]. *The translation function from* LKF *to* LK *was obtained from the translation function defined in Definition 2.8 by deleting the conditions* $f({\sim}p) := \neg p$, $f(\neg\alpha) := \neg f(\alpha)$, *and* $f({\sim}\neg\alpha) := f(\alpha)$ *and adding the condition* $f({\sim}p) := {\sim}p$. *It is remarked that the symbol* $\sim$ *was not used in* [16], *but the symbol* $\neg$ *was used as only one negation symbol.*

3. *A similar translation to the translation defined in Definition 2.8 has been used by Gurevich* [12], *Rautenberg* [21], *and Vorob'ev* [28] *to embed some variants of Nelson's constructive logics* [1, 19] *into intuitionistic logic.*

Theorem 2.10 (Weak syntactical embedding from ELK into LK). *Let* Γ *and* Δ *be (possibly empty) sets of formulas in* $\mathcal{L}_{\text{ELK}}$ *and* f *be the mapping defined in Definition 2.8.*

1. *If* $\text{ELK} \vdash \Gamma \Rightarrow \Delta$, *then* $\text{LK} \vdash f(\Gamma) \Rightarrow f(\Delta)$.

2. *If* $\text{LK} - (\text{cut}) \vdash f(\Gamma) \Rightarrow f(\Delta)$, *then* $\text{ELK} - (\text{cut}) \vdash \Gamma \Rightarrow \Delta$.

Proof.

1. By induction on the proofs P of $\Gamma \Rightarrow \Delta$ in ELK. We distinguish the cases according to the last inference of P and show only the following cases.

   (a) Case ${\sim}p \Rightarrow {\sim}p$: The last inference of P is of the form: ${\sim}p \Rightarrow {\sim}p$ for any $p \in \Phi$. We have $\text{LK} \vdash \neg p \Rightarrow \neg p$. Thus, we obtain $\text{LK} \vdash f({\sim}p) \Rightarrow f({\sim}p)$ by the definition of f. It is remarked that the translation $f({\sim}p)$ can be given for any forms of the atomic formula p (e.g., if p is of the form $p(x, y)$, then $f({\sim}p(x, y))$ is of the form $\neg p(x, y)$).

(b) Case $p, \sim p \Rightarrow$: The last inference of P is of the form: $p, \sim p \Rightarrow$ for any $p \in \Phi$. Using ($\neg$left), we obtain LK $\vdash p, \neg p \Rightarrow$. Thus, we obtain LK $\vdash f(p), f(\sim p) \Rightarrow$ by the definition of f.

(c) Case ($\sim\exists$left): The last inference of P is of the form:

$$\frac{\Gamma \Rightarrow \Delta, \sim\alpha[z/x]}{\Gamma \Rightarrow \Delta, \sim\exists x\alpha}\ (\sim\exists\text{right}).$$

By induction hypothesis, we have LK $\vdash f(\Gamma) \Rightarrow f(\Delta), f(\sim\alpha[z/x])$. Then, we obtain:

$$\frac{\begin{array}{c}\vdots\ \textit{Ind. hyp.}\\ f(\Gamma) \Rightarrow f(\Delta), f(\sim\alpha[z/x])\end{array}}{f(\Gamma) \Rightarrow f(\Delta), \forall x f(\sim\alpha)}\ (\forall\text{right})$$

where $\forall x f(\sim\alpha)$ coincides with $f(\sim\exists x\alpha)$ by the definition of f. It is remarked that f is independent of terms (i.e., f does not change the terms from the original ones). Thus, the above application of ($\forall$right) is guaranteed.

2. By induction on the proofs Q of $f(\Gamma) \Rightarrow f(\Delta)$ in LK – (cut). We distinguish the cases according to the last inference of Q and show only the following case.

   Case ($\forall$right): The last inference of Q is ($\forall$right).

   (a) Subcase (1): The last inference of Q is of the form:

   $$\frac{f(\Gamma) \Rightarrow f(\Delta), f(\alpha[z/x])}{f(\Gamma) \Rightarrow f(\Delta), f(\forall x\alpha)}\ (\forall\text{right})$$

   where $f(\forall x\alpha)$ coincides with $\forall x f(\alpha)$ by the definition of f. By induction hypothesis, we have ELK – (cut) $\vdash \Gamma \Rightarrow \Delta, \alpha[z/x]$. We thus obtain:

   $$\frac{\begin{array}{c}\vdots\ \textit{Ind. hyp.}\\ \Gamma \Rightarrow \Delta, \alpha[z/x]\end{array}}{\Gamma \Rightarrow \Delta, \forall x\alpha}\ (\forall\text{right}).$$

   (b) Subcase (2): The last inference of Q is of the form:

   $$\frac{f(\Gamma) \Rightarrow f(\Delta), f(\sim\alpha[z/x])}{f(\Gamma) \Rightarrow f(\Delta), f(\sim\exists x\alpha)}\ (\forall\text{right})$$

where $f(\sim\exists x\alpha)$ coincides with $\forall x f(\sim\alpha)$ by the definition of f. By induction hypothesis, we have ELK − (cut) $\vdash \Gamma \Rightarrow \Delta, \sim\alpha[z/x]$. We thus obtain:

$$\dfrac{\begin{array}{c}\vdots\ \textit{Ind. hyp.}\\ \Gamma \Rightarrow \Delta, \sim\alpha[z/x]\end{array}}{\Gamma \Rightarrow \Delta, \sim\exists x\alpha}\ (\sim\exists\text{right}).$$

□

Theorem 2.11 (Cut-elimination for ELK). *The rule* (cut) *is admissible in cut-free* ELK.

Proof. Suppose ELK $\vdash \Gamma \Rightarrow \Delta$. Then, we have LK $\vdash f(\Gamma) \Rightarrow f(\Delta)$ by Theorem 2.10 (1), and hence LK − (cut) $\vdash f(\Gamma) \Rightarrow f(\Delta)$ by the cut-elimination theorem for LK. By Theorem 2.10 (2), we obtain ELK − (cut) $\vdash \Gamma \Rightarrow \Delta$. □

Theorem 2.12 (Syntactical embedding from ELK into LK). *Let* Γ *and* Δ *be (possibly empty) sets of formulas in* $\mathcal{L}_{\text{ELK}}$ *and* f *be the mapping defined in Definition 2.8.*

1. ELK $\vdash \Gamma \Rightarrow \Delta$ *iff* LK $\vdash f(\Gamma) \Rightarrow f(\Delta)$.

2. ELK − (cut) $\vdash \Gamma \Rightarrow \Delta$ *iff* LK − (cut) $\vdash f(\Gamma) \Rightarrow f(\Delta)$.

Proof.

1. ($\Longrightarrow$): By Theorem 2.10 (1). ($\Longleftarrow$): Suppose LK $\vdash f(\Gamma) \Rightarrow f(\Delta)$. Then we have LK − (cut) $\vdash f(\Gamma) \Rightarrow f(\Delta)$ by the cut-elimination theorem for LK. We thus obtain ELK − (cut) $\vdash \Gamma \Rightarrow \Delta$ by Theorem 2.10 (2). Therefore, we have ELK $\vdash \Gamma \Rightarrow \Delta$.

2. ($\Longrightarrow$): Suppose ELK − (cut) $\vdash \Gamma \Rightarrow \Delta$. Then we have ELK $\vdash \Gamma \Rightarrow \Delta$. We then obtain LK $\vdash f(\Gamma) \Rightarrow f(\Delta)$ by Theorem 2.10 (1). Therefore, we obtain LK − (cut) $\vdash f(\Gamma) \Rightarrow f(\Delta)$ by the cut-elimination theorem for LK. ($\Longleftarrow$): By Theorem 2.10 (2).

□

Remark 2.13. *We make the following remarks.*

1. *A weak theorem for syntactically embedding* LKF [16] *into* LK *could not be proved in a similar way as that for Theorem 2.10.*

2. *By this situation, the embedding-based syntactical proof of the cut-elimination theorem for* LKF *was not obtained in* [16]. *Thus, the embedding-based simple syntactical proof of the cut-elimination theorem for* ELK *is considered to be a novel contribution of this study.*

3. *By Theorem 2.11, we can obtain the following facts: (1)* ELK *is a conservative extension of* LKF *and* LK*, although* $\sim$ *and* $\neg$ *are equivalent, (2) the cut-elimination theorems for* LKF *and* LK *hold, and (3)* LKF *and* LK *are the* $\{\wedge, \vee, \rightarrow, \sim\}$*- and* $\{\wedge, \vee, \rightarrow, \neg\}$*-fragments of* ELK*, respectively. Thus, an alternative embedding-based syntactical proof of the cut-elimination theorem for* LKF*, which was not obtained in* [16]*, is obtained by Theorem 2.11. This is also a new contribution of this study.*

4. *A theorem for syntactically embedding* LKF *into* LK *was proved in* [16]*, but the same item 2 as that for Theorem 2.12 was not proved in a similar way as that for Theorem 2.12. The proof of the item 1 in the theorem for embedding* LKF *into* LK *was required to use the cut rule in* LKF*. Thus, the direct proof of (the item 2 of) the strong theorem for embedding* ELK *into* LK *is also considered to be a novel contribution of this study.*

2.2 Sequent calculus for Gurevich logic

Next, we present a Gentzen-style sequent calculus ELJ [15] for GL. The language of ELJ is the same as that of ELK, but $\neg$ and $\sim$ are used as the intuitionistic negation and strong negation connectives, respectively. The same notations and notions as those for ELK are used for ELJ. However, the notion of sequent should be modified for ELJ. An *intuitionistic sequent* (simply called *sequent*) for ELJ is an expression of the form $\Gamma \Rightarrow \gamma$ where γ is a single formula or the empty set. The same names of inference rules as those of ELK are also used for ELJ, although the forms of inference rules in ELJ are different from those in ELK.

We now introduce a Gentzen-style sequent calculus ELJ for GL.

Definition 2.14 (ELJ). *In the following definition, we use the symbol* γ *to represent an arbitrary formula or the empty set and the symbols* p*,* t*, and* z *to represent the same objects as those indicated in Definition 2.1.*

The initial sequents of ELJ *are of the form:*

$$p \Rightarrow p \qquad {\sim}p \Rightarrow {\sim}p \qquad p, {\sim}p \Rightarrow .$$

The structural inference rules of ELJ *are of the form:*

$$\frac{\Gamma \Rightarrow \alpha \quad \alpha, \Sigma \Rightarrow \gamma}{\Gamma, \Sigma \Rightarrow \gamma}\ (\text{cut}) \quad \frac{\Gamma \Rightarrow \gamma}{\alpha, \Gamma \Rightarrow \gamma}\ (\text{we-left}) \quad \frac{\Gamma \Rightarrow}{\Gamma \Rightarrow \alpha}\ (\text{we-right}).$$

The normal logical inference rules of ELJ *are of the form:*

$$\frac{\alpha, \Gamma \Rightarrow \gamma}{\alpha \wedge \beta, \Gamma \Rightarrow \gamma} \ (\wedge\text{left1}) \quad \frac{\beta, \Gamma \Rightarrow \gamma}{\alpha \wedge \beta, \Gamma \Rightarrow \gamma} \ (\wedge\text{left2})$$

$$\frac{\Gamma \Rightarrow \alpha \quad \Gamma \Rightarrow \beta}{\Gamma \Rightarrow \alpha \wedge \beta} \ (\wedge\text{right}) \quad \frac{\alpha, \Gamma \Rightarrow \gamma \quad \beta, \Gamma \Rightarrow \gamma}{\alpha \vee \beta, \Gamma \Rightarrow \gamma} \ (\vee\text{left})$$

$$\frac{\Gamma \Rightarrow \alpha}{\Gamma \Rightarrow \alpha \vee \beta} \ (\vee\text{right1}) \quad \frac{\Gamma \Rightarrow \beta}{\Gamma \Rightarrow \alpha \vee \beta} \ (\vee\text{right2})$$

$$\frac{\Gamma \Rightarrow \alpha \quad \beta, \Sigma \Rightarrow \gamma}{\alpha \rightarrow \beta, \Gamma, \Sigma \Rightarrow \gamma} \ (\rightarrow\text{left}) \quad \frac{\alpha, \Gamma \Rightarrow \beta}{\Gamma \Rightarrow \alpha \rightarrow \beta} \ (\rightarrow\text{right})$$

$$\frac{\Gamma \Rightarrow \alpha}{\neg\alpha, \Gamma \Rightarrow} \ (\neg\text{left}) \quad \frac{\alpha, \Gamma \Rightarrow}{\Gamma \Rightarrow \neg\alpha} \ (\neg\text{right})$$

$$\frac{\alpha[t/x], \Gamma \Rightarrow \gamma}{\forall x\alpha, \Gamma \Rightarrow \gamma} \ (\forall\text{left}) \quad \frac{\Gamma \Rightarrow \alpha[z/x]}{\Gamma \Rightarrow \forall x\alpha} \ (\forall\text{right})$$

$$\frac{\alpha[z/x], \Gamma \Rightarrow \gamma}{\exists x\alpha, \Gamma \Rightarrow \gamma} \ (\exists\text{left}) \quad \frac{\Gamma \Rightarrow \alpha[t/x]}{\Gamma \Rightarrow \exists x\alpha} \ (\exists\text{right}).$$

The strongly-negated logical inference rules of ELJ *are of the form:*

$$\frac{\alpha, \Gamma \Rightarrow \gamma}{{\sim}{\sim}\alpha, \Gamma \Rightarrow \gamma} \ ({\sim}{\sim}\text{left}) \quad \frac{\Gamma \Rightarrow \alpha}{\Gamma \Rightarrow {\sim}{\sim}\alpha} \ ({\sim}{\sim}\text{right})$$

$$\frac{{\sim}\alpha, \Gamma \Rightarrow \gamma \quad {\sim}\beta, \Gamma \Rightarrow \gamma}{{\sim}(\alpha \wedge \beta), \Gamma \Rightarrow \gamma} \ ({\sim}\wedge\text{left})$$

$$\frac{\Gamma \Rightarrow {\sim}\alpha}{\Gamma \Rightarrow {\sim}(\alpha \wedge \beta)} \ ({\sim}\wedge\text{right1}) \quad \frac{\Gamma \Rightarrow {\sim}\beta}{\Gamma \Rightarrow {\sim}(\alpha \wedge \beta)} \ ({\sim}\wedge\text{right2})$$

$$\frac{{\sim}\alpha, \Gamma \Rightarrow \gamma}{{\sim}(\alpha \vee \beta), \Gamma \Rightarrow \gamma} \ ({\sim}\vee\text{left1}) \quad \frac{{\sim}\beta, \Gamma \Rightarrow \gamma}{{\sim}(\alpha \vee \beta), \Gamma \Rightarrow \gamma} \ ({\sim}\vee\text{left2})$$

$$\frac{\Gamma \Rightarrow {\sim}\alpha \quad \Gamma \Rightarrow {\sim}\beta}{\Gamma \Rightarrow {\sim}(\alpha \vee \beta)} \ ({\sim}\vee\text{right})$$

$$\frac{\alpha, \Gamma \Rightarrow \gamma}{{\sim}(\alpha \rightarrow \beta), \Gamma \Rightarrow \gamma} \ ({\sim}{\rightarrow}\text{left1}) \quad \frac{{\sim}\beta, \Gamma \Rightarrow \gamma}{{\sim}(\alpha \rightarrow \beta), \Gamma \Rightarrow \gamma} \ ({\sim}{\rightarrow}\text{left2})$$

$$\frac{\Gamma \Rightarrow \alpha \quad \Gamma \Rightarrow {\sim}\beta}{\Gamma \Rightarrow {\sim}(\alpha \rightarrow \beta)} \ ({\sim}{\rightarrow}\text{right})$$

$$\frac{\alpha, \Gamma \Rightarrow \delta}{{\sim}\neg\alpha, \Gamma \Rightarrow \delta} \ ({\sim}\neg\text{left}) \quad \frac{\Gamma \Rightarrow \alpha}{\Gamma \Rightarrow {\sim}\neg\alpha} \ ({\sim}\neg\text{right})$$

$$\frac{\sim\alpha[z/x], \Gamma \Rightarrow \gamma}{\sim\forall x\alpha, \Gamma \Rightarrow \gamma} \ (\sim\forall\text{left}) \quad \frac{\Gamma \Rightarrow \sim\alpha[t/x]}{\Gamma \Rightarrow \sim\forall x\alpha} \ (\sim\forall\text{right})$$

$$\frac{\sim\alpha[t/x], \Gamma \Rightarrow \gamma}{\sim\exists x\alpha, \Gamma \Rightarrow \gamma} \ (\sim\exists\text{left}) \quad \frac{\Gamma \Rightarrow \sim\alpha[z/x]}{\Gamma \Rightarrow \sim\exists x\alpha} \ (\sim\exists\text{right}).$$

Proposition 2.15. *Sequents of the form $\alpha \Rightarrow \alpha$ for any formula α are provable in cut-free* ELJ.

Proof. By induction on α. □

Proposition 2.16. *Sequents of the form $\alpha, \sim\alpha \Rightarrow$ for any formula α are provable in cut-free* ELJ.

Proof. By induction on α. We use Proposition 2.15. □

The following proposition shows that the sequents that correspond to the axiom schemes introduced by Gurevich in [12] are provable in cut-free ELJ.

Proposition 2.17. *The following sequents are provable in cut-free* ELJ*: For any formulas α and β,*

1. $\sim\sim\alpha \Leftrightarrow \alpha$,
2. $\sim(\alpha\wedge\beta) \Leftrightarrow \sim\alpha\vee\sim\beta$,
3. $\sim(\alpha\vee\beta) \Leftrightarrow \sim\alpha\wedge\sim\beta$,
4. $\sim(\alpha\rightarrow\beta) \Leftrightarrow \alpha\wedge\sim\beta$,
5. $\sim\neg\alpha \Leftrightarrow \alpha$,
6. $\sim\forall x\alpha \Leftrightarrow \exists x\sim\alpha$,
7. $\sim\exists x\alpha \Leftrightarrow \forall x\sim\alpha$,
8. $\alpha\wedge\sim\alpha \Rightarrow \beta$,
9. $\sim\alpha \Rightarrow \neg\alpha$.

Proof. By using Propositions 2.15 and 2.16. □

The following theorem was shown in [15].

Theorem 2.18 (Cut-elimination for ELJ)**.** *The rule* (cut) *is admissible in cut-free* ELJ.

Proof. See [15]. □

Next, we introduce a Gentzen-style sequent calculus LJ for IL. The language of LJ is obtained from that of ELJ by deleting $\sim$.

Definition 2.19 (LJ). *A Gentzen-style sequent calculus* LJ *for* IL *is defined as the* $\sim$*-free part of* ELJ *(i.e., it is obtained from* ELJ *by deleting the negated initial sequents and negated logical inference rules).*

Remark 2.20. *We make the following remarks.*

1. *A Gentzen-style sequent calculus* $\overline{\mathrm{G}}$ *originally introduced by Gurevich* [12] *used the following logical inference rule for* $\sim$ *instead of the initial-like sequents of the form* $\alpha, {\sim}\alpha \Rightarrow$*:*

   $$\frac{\Gamma \Rightarrow \alpha}{{\sim}\alpha, \Gamma \Rightarrow}\ ({\sim}\text{left}).$$

   The cut-elimination theorem for $\overline{\mathrm{G}}$ *was shown in* [12].

2. *The systems* ELJ *and* $\overline{\mathrm{G}}$ *are theorem-equivalent. This fact is shown as follows. Using* (${\sim}$left)*, we can prove the sequents of the form* $\alpha, {\sim}\alpha \Rightarrow$ *in* $\overline{\mathrm{G}}$*. Conversely, we can show that* (${\sim}$left) *is derivable in* ELJ *by:*

   $$\dfrac{\begin{array}{c}\vdots \\ \Gamma \Rightarrow \alpha\end{array} \quad \begin{array}{c}\vdots\ \textit{Prop. } 2.16 \\ \alpha, {\sim}\alpha \Rightarrow\end{array}}{{\sim}\alpha, \Gamma \Rightarrow}\ (\text{cut}).$$

3. *We can assume that the original axiom scheme* ${\sim}\alpha \rightarrow \neg\alpha$ *by Gurevich can be replaced with the axiom scheme* $({\sim}\alpha \wedge \alpha) \rightarrow \beta$*. This fact is explained as follows. Let* ELJ_1 *be the system that is obtained from* ELJ *by replacing the initial sequents with the initial sequents of the form* $\alpha \Rightarrow \alpha$ *and* ${\sim}\alpha, \alpha \Rightarrow \beta$ *for any formulas* α *and* β *and let* ELJ_2 *be the system that is obtained from* ELJ_1 *by replacing the initial sequents of the form* $\alpha, {\sim}\alpha \Rightarrow \beta$ *with the initial sequents of the form* ${\sim}\alpha \Rightarrow \neg\alpha$ *for any formula* α*. Then* ELJ*,* ELJ_1*, and* ELJ_2 *are theorem-equivalent. The theorem-equivalence of* ELJ *and* ELJ_1 *is obvious. Thus, we show the theorem-equivalence of* ELJ_1 *and* ELJ_2 *by:*

$$\dfrac{\dfrac{{\sim}\alpha \Rightarrow \neg\alpha \quad \dfrac{\alpha \Rightarrow \alpha}{\neg\alpha, \alpha \Rightarrow}\ (\neg\text{left})}{{\sim}\alpha, \alpha \Rightarrow}\ (\text{cut})}{{\sim}\alpha, \alpha \Rightarrow \beta}\ (\text{we-right}) \qquad \dfrac{\dfrac{\alpha, {\sim}\alpha \Rightarrow \neg\gamma \quad \dfrac{\alpha, {\sim}\alpha \Rightarrow \gamma}{\neg\gamma, \alpha, {\sim}\alpha \Rightarrow}\ (\neg\text{left})}{\alpha, {\sim}\alpha \Rightarrow}\ (\text{cut})}{{\sim}\alpha \Rightarrow \neg\alpha}\ (\neg\text{right}).$$

4. *As mentioned previously, a Gentzen-style sequent calculus for Nelson's* N3 *is obtained from* ELJ *by deleting the logical inference rules concerning* $\neg$ *and a Gentzen-style sequent calculus for Nelson's* N4 *is obtained from* ELJ *by deleting the logical inference rules concerning* $\neg$ *and the initial sequents of the form* $p, {\sim}p \Rightarrow$. *The cut-elimination theorems for these systems for* N3 *and* N4 *hold.*

5. *The name "*LJ*" used in this paper is from Gentzen's original sequent calculus* LJ *for* IL [10]. *In this paper, the name* LJ *is used for a small modification of the original* LJ. *The same proposition as Proposition 2.15 holds for* LJ. *The cut-elimination theorem for* LJ *holds. For more information on Gentzen's* LJ, *consult e.g.,* [10, 25].

3 Extended Gödel–Gentzen translation

We introduce a translation from the formulas of ELK to those of ELJ. This translation is regarded as an extension of the well-known Gödel–Gentzen negative translation from the formulas of CL to those of IL.

Definition 3.1 (Extended Gödel–Gentzen negative translation). *Let* $\mathcal{L}$ *be the language (or the set of formulas) of* ELK *and* ELJ. *A mapping* h *from* $\mathcal{L}$ *to* $\mathcal{L}$ *is defined inductively by:*

1. *For any* $p \in \Phi$, $h(p) := \neg\neg p$,
2. $h(\alpha\sharp\beta) := h(\alpha)\sharp h(\beta)$ *where* $\sharp \in \{\rightarrow, \wedge\}$,
3. $h(\alpha\vee\beta) := \neg(\neg h(\alpha)\wedge\neg h(\beta))$,
4. $h(\sharp\alpha) := \sharp h(\alpha)$ *where* $\sharp \in \{\neg, \forall x\}$,
5. $h(\exists x\alpha) := \neg\forall x\neg h(\alpha)$,
6. *For any* $p \in \Phi$, $h({\sim}p) := {\sim}\neg\neg p$ *(i.e.,* $h({\sim}p) = {\sim}h(p)$*)*,
7. $h(\sharp\alpha) := h(\alpha)$ *where* $\sharp \in \{{\sim}{\sim}, {\sim}\neg\}$,
8. $h({\sim}(\alpha\wedge\beta)) := \neg(\neg h({\sim}\alpha)\wedge\neg h({\sim}\beta))$ *(i.e.,* $h({\sim}(\alpha\wedge\beta)) = h({\sim}\alpha\vee{\sim}\beta)$*)*,
9. $h({\sim}(\alpha\vee\beta)) := h({\sim}\alpha)\wedge h({\sim}\beta)$ *(i.e.,* $h({\sim}(\alpha\vee\beta)) = h({\sim}\alpha\wedge{\sim}\beta)$*)*,
10. $h({\sim}(\alpha\rightarrow\beta)) := h(\alpha)\wedge h({\sim}\beta)$ *(i.e.,* $h({\sim}(\alpha\rightarrow\beta)) = h(\alpha\wedge{\sim}\beta)$*)*,
11. $h({\sim}\forall x\alpha) := \neg\forall x\neg h({\sim}\alpha)$ *(i.e.,* $h({\sim}\forall x\alpha) = h(\exists x{\sim}\alpha)$*)*,

12. $h(\sim\exists x\alpha) := \forall x h(\sim\alpha)$ *(i.e.,* $h(\sim\exists x\alpha) = h(\forall x{\sim}\alpha)$*).*

Remark 3.2. *We cannot show the same embedding theorem (Theorem 3.6) based on the simple extended Gödel–Gentzen negative translation that adopts the simple condition* $h(\sim\alpha) := \sim h(\alpha)$ *for any formula* α *instead of the conditions from 6 to 12 in Definition 3.1. Actually, we cannot prove the cases for* ($\sim\vee$left) *and* ($\sim\to$left) *in Lemma 3.5.*

Lemma 3.3. *Let* h *be the mapping defined in Definition 3.1. For any formula* α, ELK $\vdash \alpha \Leftrightarrow h(\alpha)$.

Proof. By induction on α. We distinguish the cases according to the form of α and show only the cases for $\alpha \equiv \sim\beta$.

1. Case $\beta \equiv p$ where p is an atomic formula: We show ELK $\vdash \sim p \Leftrightarrow h(\sim p)$ as follows:

$$\dfrac{\dfrac{p, \sim p \Rightarrow}{\sim p \Rightarrow \neg p}\,(\neg\text{right})}{\sim p \Rightarrow \sim\neg\neg p}\,(\sim\neg\text{right}) \qquad \dfrac{\dfrac{\Rightarrow p, \sim p}{\neg p \Rightarrow \sim p}\,(\neg\text{left})}{\sim\neg\neg p \Rightarrow \sim p}\,(\sim\neg\text{left})$$

   where $\sim\neg\neg p$ coincides with $h(\sim p)$ by the definition of h.

2. Case $\beta \equiv \sim\gamma$: We show ELK $\vdash \sim\sim\gamma \Leftrightarrow h(\sim\sim\gamma)$. By induction hypothesis, we have: ELK $\vdash \gamma \Leftrightarrow h(\gamma)$ and hence obtain the required facts:

$$\dfrac{\begin{matrix}\vdots\ \textit{Ind. hyp.}\\ \gamma \Rightarrow h(\gamma)\end{matrix}}{\sim\sim\gamma \Rightarrow h(\gamma)}\,(\sim\sim\text{left}) \qquad \dfrac{\begin{matrix}\vdots\ \textit{Ind. hyp.}\\ h(\gamma) \Rightarrow \gamma\end{matrix}}{h(\gamma) \Rightarrow \sim\sim\gamma}\,(\sim\sim\text{right})$$

   where $h(\gamma)$ coincides with $h(\sim\sim\gamma)$ by the definition of h.

3. Case $\beta \equiv \gamma\wedge\delta$: We show ELK $\vdash \sim(\gamma\wedge\delta) \Leftrightarrow h(\sim(\gamma\wedge\delta))$. By induction hypothesis, we have: ELK $\vdash \sim\gamma \Leftrightarrow h(\sim\gamma)$ and ELK $\vdash \sim\delta \Leftrightarrow h(\sim\delta)$. We then obtain the required facts:

$$\dfrac{\dfrac{\dfrac{\dfrac{\dfrac{\begin{matrix}\vdots\ \textit{Ind. hyp.}\\ \sim\gamma \Rightarrow h(\sim\gamma)\end{matrix}}{\neg h(\sim\gamma), \sim\gamma \Rightarrow}\,(\neg\text{left})}{\neg h(\sim\gamma), \neg h(\sim\delta), \sim\gamma \Rightarrow}\,(\text{we-left})}{\neg h(\sim\gamma)\wedge\neg h(\sim\delta), \sim\gamma \Rightarrow}\,(\wedge\text{left}) \quad \dfrac{\dfrac{\dfrac{\begin{matrix}\vdots\ \textit{Ind. hyp.}\\ \sim\delta \Rightarrow h(\sim\delta)\end{matrix}}{\neg h(\sim\delta), \sim\delta \Rightarrow}\,(\neg\text{left})}{\neg h(\sim\gamma), \neg h(\sim\delta), \sim\delta \Rightarrow}\,(\text{we-left})}{\neg h(\sim\gamma)\wedge\neg h(\sim\delta), \sim\delta \Rightarrow}\,(\wedge\text{left})}{\neg h(\sim\gamma)\wedge\neg h(\sim\delta), \sim(\gamma\wedge\delta) \Rightarrow}\,(\sim\wedge\text{left})}{\sim(\gamma\wedge\delta) \Rightarrow \neg(\neg h(\sim\gamma)\wedge\neg h(\sim\delta))}\,(\neg\text{right})$$

$$\dfrac{\dfrac{\dfrac{\dfrac{\dfrac{\begin{array}{c}\vdots\ \textit{Ind. hyp.}\\ h(\sim\gamma)\Rightarrow\sim\gamma\end{array}}{\Rightarrow\sim\gamma,\neg h(\sim\gamma)}(\neg\text{right})}{\Rightarrow\sim\gamma,\sim\delta,\neg h(\sim\gamma)}(\text{we-right}) \quad \dfrac{\dfrac{\begin{array}{c}\vdots\ \textit{Ind. hyp.}\\ h(\sim\delta)\Rightarrow\sim\delta\end{array}}{\Rightarrow\sim\delta,\neg h(\sim\delta)}(\neg\text{right})}{\Rightarrow\sim\gamma,\sim\delta,\neg h(\sim\delta)}(\text{we-right})}{\Rightarrow\sim\gamma,\sim\delta,\neg h(\sim\gamma)\wedge\neg h(\sim\delta)}(\wedge\text{right})}{\neg(\neg h(\sim\gamma)\wedge\neg h(\sim\delta))\Rightarrow\sim\gamma,\sim\delta}(\neg\text{left})}{\neg(\neg h(\sim\gamma)\wedge\neg h(\sim\delta))\Rightarrow\sim(\gamma\wedge\delta)}(\sim\wedge\text{right})$$

where $\neg(\neg h(\sim\gamma)\wedge\neg h(\sim\delta))$ coincides with $h(\sim(\gamma\wedge\delta))$ by the definition of h.

4. Case $\beta\equiv\gamma\vee\delta$: We show ELK $\vdash\sim(\gamma\vee\delta)\Leftrightarrow h(\sim(\gamma\vee\delta))$. By induction hypothesis, we have: ELK $\vdash\sim\gamma\Leftrightarrow h(\sim\gamma)$ and ELK $\vdash\sim\delta\Leftrightarrow h(\sim\delta)$. We then obtain the required facts:

$$\dfrac{\dfrac{\dfrac{\begin{array}{c}\vdots\ \textit{Ind. hyp.}\\ \sim\gamma\Rightarrow h(\sim\gamma)\end{array}}{\sim\gamma,\sim\delta\Rightarrow h(\sim\gamma)}(\text{we-left})}{\sim(\gamma\vee\delta)\Rightarrow h(\sim\gamma)}(\sim\vee\text{left}) \quad \dfrac{\dfrac{\begin{array}{c}\vdots\ \textit{Ind. hyp.}\\ \sim\delta\Rightarrow h(\sim\delta)\end{array}}{\sim\gamma,\sim\delta\Rightarrow h(\sim\delta)}(\text{we-left})}{\sim(\gamma\vee\delta)\Rightarrow h(\sim\delta)}(\sim\vee\text{left})}{\sim(\gamma\vee\delta)\Rightarrow h(\sim\gamma)\wedge h(\sim\delta)}(\wedge\text{right})$$

$$\dfrac{\dfrac{\dfrac{\begin{array}{c}\vdots\ \textit{Ind. hyp.}\\ h(\sim\gamma)\Rightarrow\sim\gamma\end{array}}{h(\sim\gamma),h(\sim\delta)\Rightarrow\sim\gamma}(\text{we-left})}{h(\sim\gamma)\wedge h(\sim\delta)\Rightarrow\sim\gamma}(\wedge\text{left}) \quad \dfrac{\dfrac{\begin{array}{c}\vdots\ \textit{Ind. hyp.}\\ h(\sim\delta)\Rightarrow\sim\delta\end{array}}{h(\sim\gamma),h(\sim\delta)\Rightarrow\sim\delta}(\text{we-left})}{h(\sim\gamma)\wedge h(\sim\delta)\Rightarrow\sim\delta}(\wedge\text{left})}{h(\sim\gamma)\wedge h(\sim\delta)\Rightarrow\sim(\gamma\vee\delta)}(\sim\vee\text{right})$$

where $h(\sim\gamma)\wedge h(\sim\delta)$ coincides with $h(\sim(\gamma\vee\delta))$ by the definition of h.

5. Case $\beta\equiv\gamma\rightarrow\delta$: We show ELK $\vdash\sim(\gamma\rightarrow\delta)\Leftrightarrow h(\sim(\gamma\rightarrow\delta))$. By induction hypothesis, we have: ELK $\vdash\gamma\Leftrightarrow h(\gamma)$ and ELK $\vdash\sim\delta\Leftrightarrow h(\sim\delta)$. We then obtain the required facts:

$$\dfrac{\dfrac{\dfrac{\begin{array}{c}\vdots\ \textit{Ind. hyp.}\\ \gamma\Rightarrow h(\gamma)\end{array}}{\gamma,\sim\delta\Rightarrow h(\gamma)}(\text{we-left})}{\sim(\gamma\rightarrow\delta)\Rightarrow h(\gamma)}(\sim\rightarrow\text{left}) \quad \dfrac{\dfrac{\begin{array}{c}\vdots\ \textit{Ind. hyp.}\\ \sim\delta\Rightarrow h(\sim\delta)\end{array}}{\gamma,\sim\delta\Rightarrow h(\sim\delta)}(\text{we-left})}{\sim(\gamma\rightarrow\delta)\Rightarrow h(\sim\delta)}(\sim\rightarrow\text{left})}{\sim(\gamma\rightarrow\delta)\Rightarrow h(\gamma)\wedge h(\sim\delta)}(\wedge\text{right})$$

$$\dfrac{\dfrac{\dfrac{\overset{\vdots\ \textit{Ind. hyp.}}{h(\gamma)\Rightarrow\gamma}}{h(\gamma), h(\sim\delta)\Rightarrow\gamma}\ (\text{we-left})}{h(\gamma)\wedge h(\sim\delta)\Rightarrow\gamma}\ (\wedge\text{left}) \qquad \dfrac{\dfrac{\overset{\vdots\ \textit{Ind. hyp.}}{h(\sim\delta)\Rightarrow\sim\delta}}{h(\gamma), h(\sim\delta)\Rightarrow\sim\delta}\ (\text{we-left})}{h(\gamma)\wedge h(\sim\delta)\Rightarrow\sim\delta}\ (\wedge\text{left})}{h(\gamma)\wedge h(\sim\delta)\Rightarrow\sim(\gamma\to\delta)}\ (\sim\to\text{right})$$

where $h(\gamma)\wedge h(\sim\delta)$ coincides with $h(\sim(\gamma\to\delta))$ by the definition of h.

6. Case $\beta\equiv\neg\gamma$: We show ELK $\vdash\ \sim\neg\gamma \Leftrightarrow h(\sim\neg\gamma)$. By induction hypothesis, we have: ELK $\vdash \gamma \Leftrightarrow h(\gamma)$ and hence obtain the required facts:

$$\dfrac{\overset{\vdots\ \textit{Ind. hyp.}}{\gamma\Rightarrow h(\gamma)}}{\sim\neg\gamma\Rightarrow h(\gamma)}\ (\sim\neg\text{right}) \qquad \dfrac{\overset{\vdots\ \textit{Ind. hyp.}}{h(\gamma)\Rightarrow\gamma}}{h(\gamma)\Rightarrow\sim\neg\gamma}\ (\sim\neg\text{right})$$

where $h(\gamma)$ coincides with $h(\sim\neg\gamma)$ by the definition of h.

7. Case $\beta\equiv\forall x\gamma$: We show ELK $\vdash\ \sim\forall x\gamma \Leftrightarrow h(\sim\forall x\gamma)$. By induction hypothesis, we have: ELK $\vdash\ \sim\gamma[z/x] \Leftrightarrow h(\sim\gamma[z/x])$. We then obtain the required facts:

$$\dfrac{\dfrac{\dfrac{\dfrac{\overset{\vdots\ \textit{Ind. hyp.}}{\sim\gamma[z/x]\Rightarrow h(\sim\gamma[z/x])}}{\neg h(\sim\gamma[z/x]), \sim\gamma[z/x]\Rightarrow}\ (\neg\text{left})}{\forall x\neg h(\sim\gamma), \sim\gamma[z/x]\Rightarrow}\ (\forall\text{left})}{\forall x\neg h(\sim\gamma), \sim\forall x\gamma\Rightarrow}\ (\sim\forall\text{left})}{\sim\forall x\gamma\Rightarrow\neg\forall x\neg h(\sim\gamma)}\ (\neg\text{right}) \qquad \dfrac{\dfrac{\dfrac{\dfrac{\overset{\vdots\ \textit{Ind. hyp.}}{h(\sim\gamma[z/x])\Rightarrow\sim\gamma[z/x]}}{\Rightarrow\sim\gamma[z/x], \neg h(\sim\gamma[z/x])}\ (\neg\text{right})}{\Rightarrow\sim\forall x\gamma, \neg h(\sim\gamma[z/x])}\ (\sim\forall\text{right})}{\Rightarrow\sim\forall x\gamma, \forall x\neg h(\sim\gamma)}\ (\forall\text{right})}{\neg\forall x\neg h(\sim\gamma)\Rightarrow\sim\forall x\gamma}\ (\neg\text{left})$$

where $\neg\forall x\neg h(\sim\gamma)$ coincides with $h(\sim\forall x\gamma)$ by the definition of h.

8. Case $\beta\equiv\exists x\gamma$: We show ELK $\vdash\ \sim\exists x\gamma \Leftrightarrow h(\sim\exists x\gamma)$. By induction hypothesis, we have: ELK $\vdash\ \sim\gamma[z/x] \Leftrightarrow h(\sim\gamma[z/x])$. We then obtain the required facts:

$$\dfrac{\dfrac{\overset{\vdots\ \textit{Ind. hyp.}}{\sim\gamma[z/x]\Rightarrow h(\sim\gamma[z/x])}}{\sim\exists x\gamma\Rightarrow h(\sim\gamma[z/x])}\ (\sim\exists\text{left})}{\sim\exists x\gamma\Rightarrow\forall x h(\sim\gamma)}\ (\forall\text{right}) \qquad \dfrac{\dfrac{\overset{\vdots\ \textit{Ind. hyp.}}{h(\sim\gamma[z/x])\Rightarrow\sim\gamma[z/x]}}{\forall x h(\sim\gamma)\Rightarrow\sim\gamma[z/x]}\ (\forall\text{left})}{\forall x h(\sim\gamma)\Rightarrow\sim\exists x\gamma}\ (\sim\exists\text{right})$$

where $\forall x h(\sim\gamma)$ coincides with $h(\sim\exists x\gamma)$ by the definition of h.

$\square$

Lemma 3.4. *Let h be the mapping defined in Definition 3.1. For any formula α,* ELJ $\vdash \neg\neg h(\alpha) \Rightarrow h(\alpha)$.

Proof. By induction on α. We distinguish the cases according to the form of α and show only the cases for $\alpha \equiv {\sim}\beta$.

1. Case $\beta \equiv p$ where p is an atomic formula: We show ELJ $\vdash \neg\neg h({\sim}p) \Rightarrow h({\sim}p)$. We obtain the required fact:

$$\dfrac{\dfrac{\dfrac{\dfrac{\dfrac{\dfrac{p \Rightarrow p}{\neg p, p \Rightarrow}\ (\neg\text{left})}{{\sim}\neg\neg p, p \Rightarrow}\ ({\sim}\neg\text{left})}{p \Rightarrow \neg{\sim}\neg\neg p}\ (\neg\text{right})}{p, \neg\neg{\sim}\neg\neg p \Rightarrow}\ (\neg\text{left})}{\neg\neg{\sim}\neg\neg p \Rightarrow \neg p}\ (\neg\text{right})}{\neg\neg{\sim}\neg\neg p \Rightarrow {\sim}\neg\neg p}\ ({\sim}\neg\text{right})$$

where ${\sim}\neg\neg p$ coincides with $h({\sim}p)$ by the definition of h.

2. Case $\beta \equiv {\sim}\gamma$: We show ELJ $\vdash \neg\neg h({\sim}{\sim}\gamma) \Rightarrow h({\sim}{\sim}\gamma)$. By induction hypothesis, we have: ELJ $\vdash \neg\neg h(\gamma) \Rightarrow h(\gamma)$. We then obtain the required fact, because $h(\gamma)$ coincides with $h({\sim}{\sim}\gamma)$ by the definition of f.

3. Case $\beta \equiv \gamma\wedge\delta$: We show ELJ $\vdash \neg\neg h({\sim}(\gamma\wedge\delta)) \Rightarrow h({\sim}(\gamma\wedge\delta))$. We obtain the required fact:

$$\dfrac{\dfrac{\dfrac{\dfrac{\begin{array}{c}\vdots\ \textit{Prop. } 2.15\\ \neg h({\sim}\gamma)\wedge\neg h({\sim}\delta) \Rightarrow \neg h({\sim}\gamma)\wedge\neg h({\sim}\delta)\end{array}}{\neg(\neg h({\sim}\gamma)\wedge\neg h({\sim}\delta)), \neg h({\sim}\gamma)\wedge\neg h({\sim}\delta) \Rightarrow}\ (\neg\text{left})}{\neg h({\sim}\gamma)\wedge\neg h({\sim}\delta) \Rightarrow \neg\neg(\neg h({\sim}\gamma)\wedge\neg h({\sim}\delta))}\ (\neg\text{right})}{\neg h({\sim}\gamma)\wedge\neg h({\sim}\delta), \neg\neg\neg(\neg h({\sim}\gamma)\wedge\neg h({\sim}\delta)) \Rightarrow}\ (\neg\text{left})}{\neg\neg\neg(\neg h({\sim}\gamma)\wedge\neg h({\sim}\delta)) \Rightarrow \neg(\neg h({\sim}\gamma)\wedge\neg h({\sim}\delta))}\ (\neg\text{right})$$

where $\neg(\neg h({\sim}\gamma)\wedge\neg h({\sim}\delta))$ coincides with $h({\sim}(\gamma\wedge\delta))$ by the definition of h.

4. Case $\beta \equiv \gamma\vee\delta$: We show ELJ $\vdash \neg\neg h({\sim}(\gamma\vee\delta)) \Rightarrow h({\sim}(\gamma\vee\delta))$. By induction hypothesis, we have: ELJ $\vdash \neg\neg h({\sim}\gamma) \Rightarrow h({\sim}\gamma)$ and ELJ $\vdash \neg\neg h({\sim}\delta) \Rightarrow h({\sim}\delta)$. We then obtain the required fact:

$$\dfrac{\begin{array}{c}\vdots\ P_1\\ \neg\neg(h({\sim}\gamma)\wedge h({\sim}\delta)) \Rightarrow h({\sim}\gamma)\end{array} \quad \begin{array}{c}\vdots\ P_2\\ \neg\neg(h({\sim}\gamma)\wedge h(\neg\delta)) \Rightarrow h({\sim}\delta)\end{array}}{\neg\neg(h({\sim}\gamma)\wedge h({\sim}\delta)) \Rightarrow h({\sim}\gamma)\wedge h({\sim}\delta)}\ (\wedge\text{right})$$

where P_1 is of the form:

$$
\dfrac{
\dfrac{
\dfrac{
\dfrac{
\dfrac{
\dfrac{
\begin{matrix} \vdots\ \mathit{Prop.}\ 2.15 \\ h({\sim}\gamma) \Rightarrow h({\sim}\gamma) \end{matrix}
}{h({\sim}\gamma)\wedge h({\sim}\delta) \Rightarrow h({\sim}\gamma)}\;(\wedge\text{left1})
}{\neg h({\sim}\gamma), h({\sim}\gamma)\wedge h({\sim}\delta) \Rightarrow}\;(\neg\text{left})
}{\neg h({\sim}\gamma) \Rightarrow \neg(h({\sim}\gamma)\wedge h({\sim}\delta))}\;(\neg\text{right})
}{\neg h({\sim}\gamma), \neg\neg(h({\sim}\gamma)\wedge h({\sim}\delta)) \Rightarrow}\;(\neg\text{left})
}{\neg\neg(h({\sim}\gamma)\wedge h({\sim}\delta)) \Rightarrow \neg\neg h({\sim}\gamma)}\;(\neg\text{right})
\qquad
\begin{matrix} \vdots\ \mathit{Ind.\ hyp.} \\ \neg\neg h({\sim}\gamma) \Rightarrow h({\sim}\gamma) \end{matrix}
}{\neg\neg(h({\sim}\gamma)\wedge h({\sim}\delta)) \Rightarrow h({\sim}\gamma)}\;(\text{cut})
$$

and P_2 is of the form:

$$
\dfrac{
\dfrac{
\dfrac{
\dfrac{
\dfrac{
\dfrac{
\begin{matrix} \vdots\ \mathit{Prop.}\ 2.15 \\ h({\sim}\delta) \Rightarrow h({\sim}\delta) \end{matrix}
}{h({\sim}\gamma)\wedge h({\sim}\delta) \Rightarrow h({\sim}\delta)}\;(\wedge\text{left2})
}{\neg h({\sim}\delta), h({\sim}\gamma)\wedge h({\sim}\delta) \Rightarrow}\;(\neg\text{left})
}{\neg h({\sim}\delta) \Rightarrow \neg(h({\sim}\gamma)\wedge h({\sim}\delta))}\;(\neg\text{right})
}{\neg h({\sim}\delta), \neg\neg(h({\sim}\gamma)\wedge h({\sim}\delta)) \Rightarrow}\;(\neg\text{left})
}{\neg\neg(h({\sim}\gamma)\wedge h({\sim}\delta)) \Rightarrow \neg\neg h({\sim}\delta)}\;(\neg\text{right})
\qquad
\begin{matrix} \vdots\ \mathit{Ind.\ hyp.} \\ \neg\neg h({\sim}\delta) \Rightarrow h({\sim}\delta) \end{matrix}
}{\neg\neg(h({\sim}\gamma)\wedge h({\sim}\delta)) \Rightarrow h({\sim}\delta)}\;(\text{cut})
$$

where $h({\sim}\gamma)\wedge h({\sim}\delta)$ coincides with $h({\sim}(\gamma\vee\delta))$ by the definition of h.

5. Case $\beta \equiv \gamma{\rightarrow}\delta$: We show ELJ $\vdash \neg\neg h({\sim}(\gamma{\rightarrow}\delta)) \Rightarrow h({\sim}(\gamma{\rightarrow}\delta))$. By induction hypothesis, we have: ELJ $\vdash \neg\neg h(\gamma) \Rightarrow h(\gamma)$ and ELJ $\vdash \neg\neg h({\sim}\delta) \Rightarrow h({\sim}\delta)$. We then obtain the required fact:

$$
\dfrac{
\begin{matrix} \vdots\ P_1 \\ \neg\neg(h(\gamma)\wedge h({\sim}\delta)) \Rightarrow h(\gamma) \end{matrix}
\qquad
\begin{matrix} \vdots\ P_2 \\ \neg\neg(h(\gamma)\wedge h({\sim}\delta)) \Rightarrow h({\sim}\delta) \end{matrix}
}{\neg\neg(h(\gamma)\wedge h({\sim}\delta)) \Rightarrow h(\gamma)\wedge h({\sim}\delta)}\;(\wedge\text{right})
$$

where P_1 is of the form:

$$
\dfrac{
\dfrac{\dfrac{\dfrac{\dfrac{\dfrac{\begin{array}{c}\vdots\ \textit{Prop. 2.15}\\ h(\gamma)\Rightarrow h(\gamma)\end{array}}{h(\gamma)\wedge h(\sim\delta)\Rightarrow h(\gamma)}\,(\wedge\text{left1})}{h(\gamma)\wedge h(\sim\delta),\neg h(\gamma)\Rightarrow}\,(\neg\text{left})}{\neg h(\gamma)\Rightarrow\neg(h(\gamma)\wedge h(\sim\delta))}\,(\neg\text{right})}{\neg h(\gamma),\neg\neg(h(\gamma)\wedge h(\sim\delta))\Rightarrow}\,(\neg\text{left})}{\neg\neg(h(\gamma)\wedge h(\sim\delta))\Rightarrow\neg\neg h(\gamma)}\,(\neg\text{right})
\qquad
\begin{array}{c}\vdots\ \textit{Ind. hyp.}\\ \neg\neg h(\gamma)\Rightarrow h(\gamma)\end{array}
}{\neg\neg(h(\gamma)\wedge h(\sim\delta))\Rightarrow h(\gamma)}\,(\text{cut})
$$

and P_2 is of the form:

$$
\dfrac{
\dfrac{\dfrac{\dfrac{\dfrac{\dfrac{\begin{array}{c}\vdots\ \textit{Prop. 2.15}\\ h(\sim\delta)\Rightarrow h(\sim\delta)\end{array}}{h(\gamma)\wedge h(\sim\delta)\Rightarrow h(\sim\delta)}\,(\wedge\text{left2})}{h(\gamma)\wedge h(\sim\delta),\neg h(\sim\delta)\Rightarrow}\,(\neg\text{left})}{\neg h(\sim\delta)\Rightarrow\neg(h(\gamma)\wedge h(\sim\delta))}\,(\neg\text{right})}{\neg h(\sim\delta),\neg\neg(h(\gamma)\wedge h(\sim\delta))\Rightarrow}\,(\neg\text{left})}{\neg\neg(h(\gamma)\wedge h(\sim\delta))\Rightarrow\neg\neg h(\sim\delta)}\,(\neg\text{right})
\qquad
\begin{array}{c}\vdots\ \textit{Ind. hyp.}\\ \neg\neg h(\sim\delta)\Rightarrow h(\sim\delta)\end{array}
}{\neg\neg(h(\gamma)\wedge h(\sim\delta))\Rightarrow h(\sim\delta)}\,(\text{cut})
$$

where $h(\gamma)\wedge h(\sim\delta)$ coincides with $h(\sim(\gamma\to\delta))$ by the definition of h.

6. Case $\beta\equiv\neg\gamma$: We show ELJ $\vdash\neg\neg h(\sim\neg\gamma)\Rightarrow h(\sim\neg\gamma)$. By induction hypothesis, we have: ELJ $\vdash\neg\neg h(\gamma)\Rightarrow h(\gamma)$. We then obtain the required fact, because $h(\gamma)$ coincides with $h(\sim\neg\gamma)$ by the definition of f.

7. Case $\beta\equiv\forall x\gamma$: We show ELJ $\vdash\neg\neg h(\sim\forall x\gamma)\Rightarrow h(\sim\forall x\gamma)$. We obtain the required fact:

$$
\dfrac{\dfrac{\dfrac{\dfrac{\begin{array}{c}\vdots\ \textit{Prop. 2.15}\\ \forall x\neg h(\sim\gamma)\Rightarrow\forall x\neg h(\sim\gamma)\end{array}}{\forall x\neg h(\sim\gamma),\neg\forall x\neg h(\sim\gamma)\Rightarrow}\,(\neg\text{left})}{\forall x\neg h(\sim\gamma)\Rightarrow\neg\neg\forall x\neg h(\sim\gamma)}\,(\neg\text{right})}{\forall x\neg h(\sim\gamma),\neg\neg\neg\forall x\neg h(\sim\gamma)\Rightarrow}\,(\neg\text{left})}{\neg\neg\neg\forall x\neg h(\sim\gamma)\Rightarrow\neg\forall x\neg h(\sim\gamma)}\,(\neg\text{right})
$$

where $\neg\forall x\neg h(\sim\gamma)$ coincides with $h(\sim\forall x\gamma)$ by the definition of h.

8. Case $\beta \equiv \exists x\gamma$: We show ELJ $\vdash \neg\neg h({\sim}\exists x\gamma) \Rightarrow h({\sim}\exists x\gamma)$. By induction hypothesis, we have: ELJ $\vdash \neg\neg h({\sim}\gamma[z/x]) \Rightarrow h({\sim}\gamma[z/x])$. We then obtain the required fact:

$$\dfrac{\dfrac{\dfrac{\dfrac{\dfrac{\dfrac{\dfrac{\vdots\ \textit{Prop. 2.15}}{h({\sim}\gamma[z/x]) \Rightarrow h({\sim}\gamma[z/x])}}{\forall x h({\sim}\gamma) \Rightarrow h({\sim}\gamma[z/x])}\,(\forall\text{left})}{\forall x h({\sim}\gamma), \neg h({\sim}\gamma[z/x]) \Rightarrow}\,(\neg\text{left})}{\neg h({\sim}\gamma[z/x]) \Rightarrow \neg\forall x h({\sim}\gamma)}\,(\neg\text{right})}{\neg\neg\forall x h({\sim}\gamma), \neg h({\sim}\gamma[z/x]) \Rightarrow}\,(\neg\text{left})}{\neg\neg\forall x h({\sim}\gamma) \Rightarrow \neg\neg h({\sim}\gamma[z/x])}\,(\neg\text{right}) \quad \dfrac{\vdots\ \textit{Ind. hyp.}}{\neg\neg h({\sim}\gamma[z/x]) \Rightarrow h({\sim}\gamma[z/x])}}{\dfrac{\neg\neg\forall x h({\sim}\gamma) \Rightarrow h({\sim}\gamma[z/x])}{\neg\neg\forall x h({\sim}\gamma) \Rightarrow \forall x h({\sim}\gamma)}\,(\forall\text{right})}\,(\text{cut})$$

where $\forall x h({\sim}\gamma)$ coincides with $h({\sim}\exists x\gamma)$ by the definition of h.

□

Lemma 3.5. *Let h be the mapping defined in Definition 3.1. For any sequent $\Gamma \Rightarrow \Delta$, if* ELK $\vdash \Gamma \Rightarrow \Delta$, *then* ELJ $\vdash h(\Gamma), \neg h(\Delta) \Rightarrow$.

Proof. By induction on the proofs P of $\Gamma \Rightarrow \Delta$ in ELK. We distinguish the cases according to the last inference of P and show some cases.

1. Case ${\sim}p, p \Rightarrow$: The last inference of P is of the form: ${\sim}p, p \Rightarrow$ for any atomic formula p. In this case, we obtain the required fact:

$$\dfrac{\dfrac{\dfrac{\dfrac{p \Rightarrow p}{p, \neg p \Rightarrow}\,(\neg\text{left})}{\neg p \Rightarrow \neg p}\,(\neg\text{right})}{\neg p, \neg\neg p \Rightarrow}\,(\neg\text{left})}{{\sim}\neg\neg p, \neg\neg p \Rightarrow}\,({\sim}\neg\text{left})$$

where ${\sim}\neg\neg p$ and $\neg\neg p$ coincide with $h({\sim}p)$ and $h(p)$, respectively, by the definition of h.

2. Case $\Rightarrow {\sim}p, p$: The last inference of P is of the form: $\Rightarrow {\sim}p, p$ for any atomic formula p. In this case, we obtain the required fact:

$$\dfrac{\dfrac{\dfrac{\dfrac{\dfrac{\dfrac{p \Rightarrow p}{\neg p, p \Rightarrow}\,(\neg\text{left})}{p \Rightarrow \neg\neg p}\,(\neg\text{right})}{p, \neg\neg\neg p \Rightarrow}\,(\neg\text{left})}{\neg\neg\neg p \Rightarrow \neg p}\,(\neg\text{right})}{\neg\neg\neg p \Rightarrow {\sim}\neg\neg p}\,({\sim}\neg\text{right})}{\neg{\sim}\neg\neg p, \neg\neg\neg p \Rightarrow}\,(\neg\text{left})$$

where $\sim\neg\neg p$ and $\neg\neg p$ coincide with $h(\sim p)$ and $h(p)$, respectively, by the definition of h.

3. Case ($\sim\wedge$left): The last inference of P is of the form:

$$\frac{\sim\alpha, \Gamma \Rightarrow \Delta \quad \sim\beta, \Gamma \Rightarrow \Delta}{\sim(\alpha\wedge\beta), \Gamma \Rightarrow \Delta} \ (\sim\wedge\text{left}).$$

By induction hypothesis, we have: ELJ $\vdash h(\sim\alpha), h(\Gamma), \neg h(\Delta) \Rightarrow$ and ELJ $\vdash h(\sim\beta), h(\Gamma), \neg h(\Delta) \Rightarrow$. We then obtain the required fact:

$$\dfrac{\dfrac{\begin{array}{c}\vdots\ \textit{Ind. hyp.}\\ h(\sim\alpha), h(\Gamma), \neg h(\Delta) \Rightarrow\end{array}}{h(\Gamma), \neg h(\Delta) \Rightarrow \neg h(\sim\alpha)}\ (\neg\text{right}) \quad \dfrac{\begin{array}{c}\vdots\ \textit{Ind. hyp.}\\ h(\sim\beta), h(\Gamma), \neg h(\Delta) \Rightarrow\end{array}}{h(\Gamma), \neg h(\Delta) \Rightarrow \neg h(\sim\beta)}\ (\neg\text{right})}{\dfrac{h(\Gamma), \neg h(\Delta) \Rightarrow \neg h(\sim\alpha)\wedge\neg h(\sim\beta)}{\neg(\neg h(\sim\alpha)\wedge\neg h(\sim\beta)), h(\Gamma), \neg h(\Delta) \Rightarrow}\ (\neg\text{left})}\ (\wedge\text{right})$$

where $\neg(\neg h(\sim\alpha)\wedge\neg h(\sim\beta))$ coincides with $h(\sim(\alpha\wedge\beta))$ by the definition of h.

4. Case ($\sim\wedge$right): The last inference of P is of the form:

$$\frac{\Gamma \Rightarrow \Delta, \sim\alpha, \sim\beta}{\Gamma \Rightarrow \Delta, \sim(\alpha\wedge\beta)} \ (\sim\wedge\text{right}).$$

By induction hypothesis, we have: ELJ $\vdash h(\Gamma), \neg h(\Delta), \neg h(\sim\alpha), \neg h(\sim\beta) \Rightarrow$. We then obtain the required fact:

$$\dfrac{\dfrac{\begin{array}{c}\vdots\ \textit{Ind. hyp.}\\ h(\Gamma), \neg h(\Delta), \neg h(\sim\alpha), \neg h(\sim\beta) \Rightarrow\\ \vdots\ (\wedge\text{left1}), (\wedge\text{left2})\\ h(\Gamma), \neg h(\Delta), \neg h(\sim\alpha)\wedge\neg h(\sim\beta) \Rightarrow\end{array}}{h(\Gamma), \neg h(\Delta) \Rightarrow \neg(\neg h(\sim\alpha)\wedge\neg h(\sim\beta))}\ (\neg\text{right})}{h(\Gamma), \neg h(\Delta), \neg\neg(\neg h(\sim\alpha)\wedge\neg h(\sim\beta)) \Rightarrow}\ (\neg\text{left})$$

where $\neg(\neg h(\sim\alpha)\wedge\neg h(\sim\beta))$ coincides with $h(\sim(\alpha\wedge\beta))$ by the definition of h.

5. Case ($\sim\vee$left): The last inference of P is of the form:

$$\frac{\sim\alpha, \sim\beta, \Gamma \Rightarrow \Delta}{\sim(\alpha\vee\beta), \Gamma \Rightarrow \Delta} \ (\sim\vee\text{left}).$$

By induction hypothesis, we have: ELJ $\vdash h(\sim\alpha), h(\sim\beta), h(\Gamma), \neg h(\Delta) \Rightarrow$. We then obtain the required fact:

$$\dfrac{\dfrac{\vdots\ \textit{Ind. hyp.}}{h(\sim\alpha), h(\sim\beta), h(\Gamma), \neg h(\Delta) \Rightarrow}}{h(\sim\alpha)\wedge h(\sim\beta), h(\Gamma), \neg h(\Delta) \Rightarrow}\ \vdots\ (\wedge\text{left1}), (\wedge\text{left2})$$

where $h(\sim\alpha)\wedge h(\sim\beta)$ coincides with $h(\sim(\alpha\vee\beta))$ by the definition of h.

6. Case ($\sim\vee$right): The last inference of P is of the form:

$$\frac{\Gamma \Rightarrow \Delta, \sim\alpha \quad \Gamma \Rightarrow \Delta, \sim\beta}{\Gamma \Rightarrow \Delta, \sim(\alpha\vee\beta)}\ (\sim\vee\text{right}).$$

By induction hypothesis, we have: ELJ $\vdash h(\Gamma), \neg h(\Delta), \neg h(\sim\alpha) \Rightarrow$ and ELJ $\vdash$ $h(\Gamma), \neg h(\Delta), \neg h(\sim\beta) \Rightarrow$. By Lemma 3.4, we have: ELJ $\vdash \neg\neg h(\sim\alpha) \Rightarrow h(\sim\alpha)$ and ELJ $\vdash \neg\neg h(\sim\beta) \Rightarrow h(\sim\beta)$. We then obtain the required fact:

$$\dfrac{\dfrac{\begin{array}{c}\vdots\ P_1 \\ h(\Gamma), \neg h(\Delta) \Rightarrow h(\sim\alpha)\end{array} \quad \begin{array}{c}\vdots\ P_2 \\ h(\Gamma), \neg h(\Delta) \Rightarrow h(\sim\beta)\end{array}}{h(\Gamma), \neg h(\Delta) \Rightarrow h(\sim\alpha)\wedge h(\sim\beta)}\ (\wedge\text{right})}{h(\Gamma), \neg h(\Delta), \neg(h(\sim\alpha)\wedge h(\sim\beta)) \Rightarrow}\ (\neg\text{left})$$

where P_1 is of the form:

$$\dfrac{\dfrac{\begin{array}{c}\vdots\ \textit{Ind. hyp.} \\ h(\Gamma), \neg h(\Delta), \neg h(\sim\alpha) \Rightarrow\end{array}}{h(\Gamma), \neg h(\Delta) \Rightarrow \neg\neg h(\sim\alpha)}\ (\neg\text{right}) \quad \begin{array}{c}\vdots\ \textit{Lemma } 3.4 \\ \neg\neg h(\sim\alpha) \Rightarrow h(\sim\alpha)\end{array}}{h(\Gamma), \neg h(\Delta) \Rightarrow h(\sim\alpha)}\ (\text{cut})$$

and P_2 is of the form:

$$\dfrac{\dfrac{\begin{array}{c}\vdots\ \textit{Ind. hyp.} \\ h(\Gamma), \neg h(\Delta), \neg h(\sim\beta) \Rightarrow\end{array}}{h(\Gamma), \neg h(\Delta) \Rightarrow \neg\neg h(\sim\beta)}\ (\neg\text{right}) \quad \begin{array}{c}\vdots\ \textit{Lemma } 3.4 \\ \neg\neg h(\sim\beta) \Rightarrow h(\sim\beta)\end{array}}{h(\Gamma), \neg h(\Delta) \Rightarrow h(\sim\beta)}\ (\text{cut})$$

where $h(\sim\alpha)\wedge h(\sim\beta)$ coincides with $h(\sim(\alpha\vee\beta))$ by the definition of h.

7. Case ($\sim\!\to$left): The last inference of P is of the form:

$$\frac{\alpha, {\sim}\beta, \Gamma \Rightarrow \Delta}{{\sim}(\alpha\to\beta), \Gamma \Rightarrow \Delta}\ ({\sim}\!\to\text{left}).$$

By induction hypothesis, we have: ELJ $\vdash h(\alpha), h({\sim}\beta), h(\Gamma), \neg h(\Delta) \Rightarrow$. We then obtain the required fact:

$$\begin{array}{c} \vdots\ \textit{Ind. hyp.} \\ h(\alpha), h({\sim}\beta), h(\Gamma), \neg h(\Delta) \Rightarrow \\ \vdots\ (\wedge\text{left1}), (\wedge\text{left2}) \\ h(\alpha)\wedge h({\sim}\beta), h(\Gamma), \neg h(\Delta) \Rightarrow \end{array}$$

where $h(\alpha)\wedge h({\sim}\beta)$ coincides with $h({\sim}(\alpha\to\beta))$ by the definition of h.

8. Case ($\sim\!\to$right): The last inference of P is of the form:

$$\frac{\Gamma \Rightarrow \Delta, \alpha \quad \Gamma \Rightarrow \Delta, {\sim}\beta}{\Gamma \Rightarrow \Delta, {\sim}(\alpha\to\beta)}\ ({\sim}\!\to\text{right}).$$

By induction hypothesis, we have: ELJ $\vdash h(\Gamma), \neg h(\Delta), \neg h(\alpha) \Rightarrow$ and ELJ $\vdash h(\Gamma), \neg h(\Delta), \neg h({\sim}\beta) \Rightarrow$. By Lemma 3.4, we have: ELJ $\vdash \neg\neg h(\alpha) \Rightarrow h(\alpha)$ and ELJ $\vdash \neg\neg h({\sim}\beta) \Rightarrow h({\sim}\beta)$. We then obtain the required fact:

$$\dfrac{\dfrac{\begin{array}{c}\vdots\ P_1 \\ h(\Gamma), \neg h(\Delta) \Rightarrow h(\alpha)\end{array} \quad \begin{array}{c}\vdots\ P_2 \\ h(\Gamma), \neg h(\Delta) \Rightarrow h({\sim}\beta)\end{array}}{h(\Gamma), \neg h(\Delta) \Rightarrow h(\alpha)\wedge h({\sim}\beta)}\ (\wedge\text{right})}{h(\Gamma), \neg h(\Delta), \neg(h(\alpha)\wedge h({\sim}\beta)) \Rightarrow}\ (\neg\text{left})$$

where P_1 is of the form:

$$\dfrac{\dfrac{\begin{array}{c}\vdots\ \textit{Ind. hyp.} \\ h(\Gamma), \neg h(\Delta), \neg h(\alpha) \Rightarrow\end{array}}{h(\Gamma), \neg h(\Delta) \Rightarrow \neg\neg h(\alpha)}\ (\neg\text{right}) \quad \begin{array}{c}\vdots\ \textit{Lemma } 3.4 \\ \neg\neg h(\alpha) \Rightarrow h(\alpha)\end{array}}{h(\Gamma), \neg h(\Delta) \Rightarrow h(\alpha)}\ (\text{cut})$$

and P_2 is of the form:

$$\dfrac{\dfrac{\begin{array}{c}\vdots\ \textit{Ind. hyp.} \\ h(\Gamma), \neg h(\Delta), \neg h({\sim}\beta) \Rightarrow\end{array}}{h(\Gamma), \neg h(\Delta) \Rightarrow \neg\neg h({\sim}\beta)}\ (\neg\text{right}) \quad \begin{array}{c}\vdots\ \textit{Lemma } 3.4 \\ \neg\neg h({\sim}\beta) \Rightarrow h({\sim}\beta)\end{array}}{h(\Gamma), \neg h(\Delta) \Rightarrow h({\sim}\beta)}\ (\text{cut})$$

where $h(\alpha)\wedge h({\sim}\beta)$ coincides with $h({\sim}(\alpha\to\beta))$ by the definition of h.

9. Case ($\sim\neg$right): The last inference of P is of the form:

$$\frac{\Gamma \Rightarrow \Delta, \alpha}{\Gamma \Rightarrow \Delta, \sim\neg\alpha} \ (\sim\neg\text{right}).$$

By induction hypothesis, we have: ELJ $\vdash h(\Gamma), \neg h(\Delta), \neg h(\alpha) \Rightarrow$. We then obtain the required fact ELJ $\vdash h(\Gamma), \neg h(\Delta), \neg h(\sim\neg\alpha) \Rightarrow$, because $h(\sim\neg\alpha)$ coincides with $h(\alpha)$ by the definition of h.

10. Case ($\sim\forall$left): The last inference of P is of the form:

$$\frac{\sim\alpha[z/x], \Gamma \Rightarrow \Delta}{\sim\forall x\alpha, \Gamma \Rightarrow \Delta} \ (\sim\forall\text{left}).$$

By induction hypothesis, we have: ELJ $\vdash h(\sim\alpha[z/x]), h(\Gamma), \neg h(\Delta) \Rightarrow$. We then obtain the required fact:

$$\dfrac{\dfrac{\dfrac{\vdots \ \textit{Ind. hyp.}}{h(\sim\alpha[z/x]), h(\Gamma), \neg h(\Delta) \Rightarrow}}{h(\Gamma), \neg h(\Delta) \Rightarrow \neg h(\sim\alpha[z/x])} \ (\neg\text{right})}{\dfrac{h(\Gamma), \neg h(\Delta) \Rightarrow \forall x\neg h(\sim\alpha)}{\neg\forall x\neg h(\sim\alpha), h(\Gamma), \neg h(\Delta) \Rightarrow} \ (\neg\text{left})} \ (\forall\text{right})$$

where $\neg\forall x\neg h(\sim\alpha)$ coincides with $h(\sim\forall x\alpha)$ by the definition of h.

11. Case ($\sim\forall$right): The last inference of P is of the form:

$$\frac{\Gamma \Rightarrow \Delta, \sim\alpha[t/x]}{\Gamma \Rightarrow \Delta, \sim\forall x\alpha} \ (\sim\forall\text{right}).$$

By induction hypothesis, we have: ELJ $\vdash h(\Gamma), \neg h(\Delta), \neg h(\sim\alpha[t/x]) \Rightarrow$. We then obtain the required fact:

$$\begin{array}{c} \vdots \ \textit{Ind. hyp.} \\ h(\Gamma), \neg h(\Delta), \neg h(\sim\alpha[t/x]) \Rightarrow \\ \hline h(\Gamma), \neg h(\Delta), \forall x\neg h(\sim\alpha) \Rightarrow \\ \hline h(\Gamma), \neg h(\Delta) \Rightarrow \neg\forall x\neg h(\sim\alpha) \\ \hline h(\Gamma), \neg h(\Delta), \neg\neg\forall x\neg h(\sim\alpha) \Rightarrow \end{array} \begin{array}{l} \\ (\forall\text{left}) \\ (\neg\text{right}) \\ (\neg\text{left}) \end{array}$$

where $\neg\forall x\neg h(\sim\alpha)$ coincides with $h(\sim\forall x\alpha)$ by the definition of h.

12. Case ($\sim\exists$left): The last inference of P is of the form:

$$\dfrac{\sim\alpha[t/x], \Gamma \Rightarrow \Delta}{\sim\exists x\alpha, \Gamma \Rightarrow \Delta}\ (\sim\exists\text{left}).$$

By induction hypothesis, we have: $\mathrm{ELJ} \vdash h(\sim\alpha[t/x]), h(\Gamma), \neg h(\Delta) \Rightarrow$. We obtain the required fact:

$$\dfrac{\begin{array}{c}\vdots\ \textit{Ind. hyp.}\\ h(\sim\alpha[t/x]), h(\Gamma), \neg h(\Delta) \Rightarrow\end{array}}{\forall x h(\sim\alpha), h(\Gamma), \neg h(\Delta) \Rightarrow}\ (\forall\text{left})$$

where $\forall x h(\sim\alpha)$ coincides with $h(\sim\exists x\alpha)$ by the definition of h.

13. Case ($\sim\exists$right): The last inference of P is of the form:

$$\dfrac{\Gamma \Rightarrow \Delta, \sim\alpha[z/x]}{\Gamma \Rightarrow \Delta, \sim\exists x\alpha}\ (\sim\exists\text{right}).$$

By induction hypothesis, we have: $\mathrm{ELJ} \vdash h(\Gamma), \neg h(\Delta), \neg h(\sim\alpha[z/x]) \Rightarrow$. By Lemma 3.4, we have: $\mathrm{ELJ} \vdash \neg\neg h(\sim\alpha[z/x]) \Rightarrow h(\sim\alpha[z/x])$. We then obtain the required fact:

$$\dfrac{\dfrac{\dfrac{\begin{array}{c}\vdots\ \textit{Ind. hyp.}\\ h(\Gamma), \neg h(\Delta), \neg h(\sim\alpha[z/x]) \Rightarrow\end{array}}{h(\Gamma), \neg h(\Delta) \Rightarrow \neg\neg h(\sim\alpha[z/x])}\ (\neg\text{right}) \qquad \begin{array}{c}\vdots\ \textit{Lemma 3.4}\\ \neg\neg h(\sim\alpha[z/x]) \Rightarrow h(\sim\alpha[z/x])\end{array}}{h(\Gamma), \neg h(\Delta) \Rightarrow h(\sim\alpha[z/x])}\ (\text{cut})}{h(\Gamma), \neg h(\Delta) \Rightarrow \forall x h(\sim\alpha)}\ (\forall\text{right})$$

where $\forall x h(\sim\alpha)$ coincides with $h(\sim\exists x\alpha)$ by the definition of h.

□

Theorem 3.6 (Gödel–Gentzen embedding from ELK into ELJ). *Let h be the mapping defined in Definition 3.1. For any formula α,* $\mathrm{ELK} \vdash \Rightarrow \alpha$ *iff* $\mathrm{ELJ} \vdash \Rightarrow h(\alpha)$.

Proof. ($\Longrightarrow$): Suppose $\mathrm{ELK} \vdash\Rightarrow \alpha$. Then, we have $\mathrm{ELJ} \vdash \neg h(\alpha) \Rightarrow$ by Lemma 3.5. By Lemma 3.4, we have $\mathrm{ELJ} \vdash \neg\neg h(\alpha) \Rightarrow h(\alpha)$. We thus obtain the required fact:

$$\dfrac{\dfrac{\begin{array}{c}\vdots\ \textit{Lemma 3.5}\\ \neg h(\alpha) \Rightarrow\end{array}}{\Rightarrow \neg\neg h(\alpha)}\ (\neg\text{right}) \qquad \begin{array}{c}\vdots\ \textit{Lemma 3.4}\\ \neg\neg h(\alpha) \Rightarrow h(\alpha)\end{array}}{\Rightarrow h(\alpha)}\ (\text{cut}).$$

($\Longleftarrow$): Suppose ELJ $\vdash \Rightarrow h(\alpha)$. Then, we have ELK $\vdash \Rightarrow h(\alpha)$ and hence obtain ELK $\vdash \Rightarrow \alpha$ by Lemma 3.3 with (cut):

$$\dfrac{\begin{array}{c}\vdots\ Hyp. \\ \Rightarrow h(\alpha)\end{array} \qquad \begin{array}{c}\vdots\ Lemma\ 3.3 \\ h(\alpha) \Rightarrow \alpha\end{array}}{\Rightarrow \alpha}\ \text{(cut)}.$$

□

We can show a theorem for embedding ELK into LJ using a slightly modified version of the translation defined in Definition 3.1.

Definition 3.7 (Modified extended Gödel–Gentzen negative translation). *Let $\mathcal{L}$ be the language (or the set of formulas) of* ELK *and $\mathcal{L}^-$ be the language (or the set of formulas) of* LJ *(i.e., it is obtained from $\mathcal{L}$ by deleting $\sim$). A mapping h from $\mathcal{L}$ to $\mathcal{L}^-$ is obtained from the conditions in Definition 3.1 by replacing the condition 6 with the following condition:*

6′. *For any $p \in \Phi$, $h({\sim}p) := \neg h(p)$ (i.e., $h({\sim}p) = \neg\neg\neg p$).*

Remark 3.8. *We can prove the same lemmas with respect to Definition 3.7 as Lemmas 3.3, 3.4, and 3.5. The same lemma as Lemma 3.3 can be proved using the same manner as that of Lemma 3.3. The same lemmas as Lemmas 3.4 and 3.5 can be proved using* ($\neg$left) *and* ($\neg$right) *instead of* (${\sim}\neg$left) *and* (${\sim}\neg$right).

Theorem 3.9 (Gödel–Gentzen embedding from ELK into LJ). *Let h be the mapping defined in Definition 3.7. For any formula α,* ELK $\vdash \Rightarrow \alpha$ *iff* LJ $\vdash \Rightarrow h(\alpha)$.

Proof. Similar to the proof of Theorem 3.6. □

4 Concluding remarks

In this study, we proved the theorems for embedding first-order classical logic (CL) into Gurevich's extended first-order intuitionistic logic with strong negation (GL) [12] and first-order intuitionistic logic (IL). To prove the embedding theorems, we first introduced the new alternative cut-free Gentzen-style sequent calculus ELK for CL by extending Gentzen's sequent calculus LK [10] for CL. Actually, ELK was obtained from LK by adding the strongly-negated initial sequents and strongly-negated logical inference rules for the strong negation connective $\sim$. Strong negation was originally introduced by Nelson in [19] and traditionally used in GL and related constructive logics [1, 19, 21, 28, 26, 30]. Next, we introduced an extended version of the Gödel–Gentzen negative translation [11, 9] from CL to IL. Then, we proved the theorem for embedding ELK into a Gentzen-style sequent calculus ELJ [15] for GL by using the extended Gödel–Gentzen negative translation (i.e., we obtained the theorem for embedding CL into GL). Finally, we proved the theorem for embedding ELK into Gentzen's sequent calculus LJ [10] for IL by using a slightly modified version of the extended Gödel–Gentzen negative translation (i.e., we obtained the alternative theorem for embedding CL into IL). By these results, the role of strong negation in CL and GL was clarified. More specifically, it was shown in this study that strong negation is an essential and natural component of CL for representing falsification-aware reasoning and plays a crucial role in smoothly proving the theorems for embedding CL into GL and IL (and some of the related theorems for CL).

We now present some related works on important GL subsystems with strong negation. A major GL subsystem (or fragment) is its propositional fragment. Spinks and Veroff [23, 24] showed that the propositional fragment of GL, called N by them, is definitionally equivalent to a certain axiomatic extension $\mathrm{NFL_{ew}}$ of the substructural logic $\mathrm{FL_{ew}}$. Another major GL subsystem is Nelson's first-order constructive three-valued logic (N3) [1, 19], which is obtained from GL by deleting intuitionistic negation (i.e., N3 is the intuitionistic-negation-less fragment of GL). Thus, a Hilbert-style axiomatic system for N3 is obtained from the axiomatic system previously presented for GL by deleting the axiom schemes concerning $\neg$. Other important GL subsystems are Nelson's first-order constructive paraconsistent four-valued logic (N4) [1, 19] and Belnap–Dunn logic (BD) (also referred to as Belnap and Dunn's useful four-valued logic, Dunn–Belnap logic, or first-degree entailment logic) [4, 5, 7]. A Hilbert-style axiomatic system for N4 is obtained from that for N3 by deleting the axiom scheme $(\alpha\wedge{\sim}\alpha)\to\beta$, and N4 is regarded as an extension of BD obtained by adding implication. Odintsov [20] showed that (propositional) N3 can be faithfully embedded into (propositional) N4. Some neighbors of Nelson logics N3 and N4 have also been studied [21, 28, 26]. Although N4 and its classical variants have been well studied, N3 and GL have not been studied extensively. For more information on N4 and its variants, see [30, 17, 18] and the references therein.

Next, we present some remarks on strong negation and its applications. As mentioned previously, ${\sim}\alpha\to\neg\alpha$, where $\neg$ is the intuitionistic negation connective, is an axiom scheme of GL. This means that $\sim$ is stronger than $\neg$ in GL. On the one hand, this fact is a reason why $\sim$ is referred to as strong negation. On the other hand, $\sim$ in N4 is not stronger than $\neg$. Actually, $\sim$ and $\neg$ are incomparable in N4. However, the negations in N4 and its variants are also referred to as strong negations. In addition, the negations in GL, N3, N4, and their variants are also refereed to as Nelson negations. Strong or Nelson negations are known to be useful for appropriately handling inexact predicates and constructive reasoning [1, 19] and are also known to be useful for handling inconsistency-tolerant (paraconsistent) reasoning in some paraconsistent subsystems of GL. Concerned with the constructive reasoning, GL, N3, and N4 have so-called the constructible falsity property: If $\sim(\alpha \wedge \beta)$ is provable in these logics, then either $\sim\alpha$ or $\sim\beta$ is provable in the logics. This property is regarded as the dual to the well-known disjunction property: If $\alpha \vee \beta$ is provable in a logic, then either α or β is provable in the logic. Concerned with the paraconsistent reasoning, N4 is a well-known paraconsistent (or inconsistency-tolerant) logic because N4 rejects the axiom scheme $(\alpha \wedge \sim\alpha) \rightarrow \beta$ (cf., a paraconsistent logic is defined as a logic that rejects $(\alpha \wedge \sim\alpha) \rightarrow \beta$). By these good characteristic properties of strong negation (i.e., the properties for representing inexact predicates, constructive reasoning, and inconsistency-tolerant reasoning), logic programming with strong negation, which is based on constructive logics with strong negation, was studied, for example, in [29, 14].

Finally, we present some related works on Gödel–Gentzen negative translation and alternative sequent calculi for CL. A comprehensive investigation of the existing negative translations including the Gödel–Gentzen translation was obtained by Ferreira and Oliva in [8], wherein the relationship among various negative translations was clarified based on the notion of modular simplification. Various alternative sequent calculi for CL and its fragments have also been studied extensively in order to obtain some computational interpretations concerning typed λ-calculi and natural deduction systems for CL and its fragments. For this direction of research, various classical sequent calculi with labeled (or annotated) formulas have been introduced and investigated, for example, in [22, 27, 13, 6, 31].

References

[1] A. Almukdad and D. Nelson, Constructible falsity and inexact predicates, Journal of Symbolic Logic 49 (1), pp. 231-233, 1984.

[2] O. Arieli and A. Avron, Reasoning with logical bilattices, Journal of Logic, Language and Information 5, pp. 25-63, 1996.

[3] O. Arieli and A. Avron, The value of the four values, Artificial Intelligence 102 (1), pp. 97-141, 1998.

[4] N.D. Belnap, A useful four-valued logic, In: Modern Uses of Multiple-Valued Logic, G. Epstein and J. M. Dunn (eds.), Dordrecht: Reidel, pp. 5-37, 1977.

[5] N.D. Belnap, How a computer should think, In: Contemporary Aspects of Philosophy, G. Ryle (ed.), Oriel Press, Stocksfield, pp. 30-56, 1977.

[6] V. Danos, J.-B. Joinet, and H. Schellinx, Computational isomorphisms in classical logic, Theoretical Computer Science 294 (3), pp. 353-378, 2003.

[7] J.M. Dunn, Intuitive semantics for first-degree entailment and 'coupled trees', Philosophical Studies 29 (3), pp. 149-168, 1976.

[8] G. Ferreira and P. Oliva, On various negative translations, Proceedings of the 3rd International Workshop on Classical Logic and Computation (CL&C 2010), EPTCS 47, pp. 21-33, 2010.

[9] G. Gentzen, Die Widerspruchsfreiheit der reinen Zahlentheorie. Mathematische Annalen 112, pp. 493-565, 1936.

[10] G. Gentzen, Collected papers of Gerhard Gentzen, M.E. Szabo (ed.), Studies in logic and the foundations of mathematics, North-Holland (English translation), 1969.

[11] K. Gödel, Zur intuitionistischen Arithmetik und Zahlentheorie. Ergebnisse eines Mathematischen Kolloquiums 4, pp. 34-38, 1933.

[12] Y. Gurevich, Intuitionistic logic with strong negation, Studia Logica 36, pp. 49-59, 1977.

[13] J.-B. Joinet, H. Schellinx, and L.T. de Falco, SN and CR for free-style $\mathrm{LK}^{\mathrm{tq}}$: Linear decorations and simulation of normalization, Journal of Symbolic Logic 67 (1), pp. 162-196, 2002.

[14] N. Kamide, A uniform proof-theoretic foundation for abstract paraconsistent logic programming, Journal of Functional and Logic Programming 2007 (1), pp. 1-36, 2007.

[15] N. Kamide, Cut-elimination, completeness, and Craig interpolation theorems for Gurevich's extended first-order intuitionistic logic with strong negation, Journal of Applied Logics 8 (5), pp. 1101-1122, 2021.

[16] N. Kamide, Falsification-aware semantics and sequent calculi for classical logic, Journal of Philosophical Logic 51 (1), pp. 99-126, 2022.

[17] N. Kamide and H. Wansing, Proof theory of Nelson's paraconsistent logic: A uniform perspective, Theoretical Computer Science 415, pp. 1-38, 2012.

[18] N. Kamide and H. Wansing, Proof theory of N4-related paraconsistent logics, Studies in Logic, Volume 54, College Publications, pp. 1-401, 2015.

[19] D. Nelson, Constructible falsity, Journal of Symbolic Logic 14, pp. 16-26, 1949.

[20] S.P. Odintsov, On the embedding of Nelson's logics, Bulletin of the Section of Logic 31(4), pp. 241-248, 2002.

[21] W. Rautenberg, Klassische und nicht-klassische Aussagenlogik, Vieweg, Braunschweig, 1979.

[22] J. E. Santo, Revisiting the correspondence between cut elimination and normalisation, Proceedings of the 27th International Colloquium on Automata, Languages and Programming (ICALP 2000), Lecture Notes in Computer Science 1853, pp. 600-611, 2000.

[23] M. Spinks and R. Veroff, Constructive logic with strong negation is a substructural logic. I, Studia Logica 88 (3), pp. 325-348, 2008.

[24] M. Spinks and R. Veroff, Constructive logic with strong negation is a substructural logic. II, Studia Logica 89 (3), pp. 401-425, 2008.

[25] G. Takeuti, Proof theory (second edition), Dover Publications, Inc. Mineola, New York, 2013.

[26] R.H. Thomason, A semantical study of constructible falsity, Zeitschrift für Mathematische Logik und Grundlagen der Mathematik 15, pp. 247-257, 1969.

[27] C. Urban and G.M. Bierman, Strong normalisation of cut-elimination in classical logic, Fundamenta Informaticae 45 (1-2), pp. 123-155, 2001.

[28] N.N. Vorob'ev, A constructive propositional calculus with strong negation (in Russian), Doklady Akademii Nauk SSSR 85, pp. 465-468, 1952.

[29] G. Wagner, Logic programming with strong negation and inexact predicates, Journal of Logic and Computation 1 (6), pp. 835-859, 1991.

[30] H. Wansing, The logic of information structures, Lecture Notes in Artificial Intelligence 681, 163 pages, 1993.

[31] D. Zunic, Computing with sequents and diagrams in classical logic - calculi *X, dX and ©X, Phd thesis, Ecole normale superieure de Lyon, France, 2007.

Received

On Weak Bases for Boolean Relational Clones and Reductions for Computational Problems

Mike Behrisch*
Institut für Diskrete Mathematik und Geometrie, TU Wien
Institut für Algebra, JKU Linz
behrisch@logic.at

Abstract

We improve an existence condition for weak bases of relational clones on finite sets. Moreover, we provide a set of singleton weak bases of Boolean relational clones different than those exhibited by Lagerkvist in [24]. We treat groups of 'similar' Boolean clones in a uniform manner with the goal of thereby simplifying proofs working by case distinction along the clones in Post's lattice. We then present relationships between weak base relations along the covering edges in Post's lattice, which can (with one exception) be exploited to obtain parsimonious reductions of computational problems related to constraint satisfaction, in which the size of the instance only grows linearly. We also investigate how the number of variables changes between instances in these reductions.

Keywords: relational clone, weak base, strong partial clone, Boolean clone, Boolean relational clone, Boolean co-clone, parsimonious reduction

1 Introduction

A fundamental problem frequently encountered throughout applications, the sciences, but also within mathematics itself is undeniably the task to solve systems of equations. In many contexts the functions appearing in such systems are restricted

This is an extended version of the article [3] that appeared in the proceedings of ISMVL 2022.

We acknowledge the work of the unknown referee, who provided several helpful remarks that lead to an improvement of the article, for example, to the inclusion of Figure 3.

*The research of M. Behrisch was partly funded by the Austrian Science Fund (FWF) through project P 33878 'Equations in Universal Algebra'.

to a special form, e.g., linear, non-linear, disjunctive, conjunctive, etc., and the nature of the functions very much determines the solution methods and the difficulty of the problem. In fact, there are two basic questions associated with a system of equations: one is to determine whether it has solutions at all, another one is to find at least their exact number (or even the solutions themselves). The first problem is known as a decision problem, deciding the existence of solutions; the second one is a counting problem, determining not only whether the set of solutions is non-empty, but also its precise cardinality.

Since functions can be understood as relations, systems of equations can be treated as constraint satisfaction problems (CSPs), i.e., questions to find satisfying variable assignments to finite conjunctions of atomic predicates corresponding to a given (usually finite) set of finitary relations—the constraint language or template of the CSP. Thanks to the CSP Dichotomy Theorem [39, 40, 11, 41], the complexity of deciding the solvability of CSPs is now finally understood on finite sets. Using the algebraic approach to CSP, mediated by the Pol-Inv Galois connection between clones and relational clones, it could be shown that deciding the existence of solutions for a CSP is always (assuming $\mathsf{P} \neq \mathsf{NP}$, either) in P or NP-complete. Seen from a fundamental point of view, a crucial fact that made this complexity classification possible was the (a priori) compatibility of the decision version of CSP with existential quantification and the equality predicate. That is to say, it was possible—before proving the Dichotomy Theorem—to show that CSPs whose parametrising relations can be expressed in terms of some other template by a definition involving just existential quantification, conjunction and equality, can be reduced in logarithmic space to the CSP given by that template. The existence of such reductions is the key fact ensuring that the dividing line between problems in P and NP-complete problems could be phrased in terms of clones and relational clones; namely, a CSP on a finite set is in P if (under $\mathsf{P} \neq \mathsf{NP}$ also only if) the template has a so-called Taylor polymorphism. As this compatibility with polymorphism clones (on the side of templates with existential quantification and equality) was known beforehand, it was also instrumental in the proof of the dichotomy.

Using Turing reductions the counting version of CSP was also shown to be (a priori) compatible with Pol-Inv [12, Theorems 2,3], and a dichotomy was proved in [9, 10], using congruence singularity as the distinguishing property, cf. [10, Theorem 2.22]. This result was subsequently characterised differently in terms of strongly balanced templates [16, Theorem 28], making the dichotomy decidable. However, when using parsimonious reductions, where—in contrast to Turing reductions—one may only reduce problems to others with the same number of solutions, or when requiring that the size of the instance not increase too much in a reduction, counting solutions of a CSP is not compatible with the introduction of existential quantification any more. In those cases and also in the analysis of enumerating solutions [37, 36] different methods are needed, as the framework provided by relational clones is too coarse to appropriately reflect the complexity. In fact, changing the allowed complexity reductions for a computational problem, for example, considering counting CSPs under parsimonious reductions, or restricting the available reductions to obtain more fine-grained complexity classifications for CSPs and related problems [22, 25, 19, 20, 28, 23, 21, 13], is a typical situation where compatibility with existential quantification and/or equality may be lost or not available. Moreover, there are several other questions in computer and information science that do not exhibit a known a priori compatibility with $\exists$, see, e.g., optimisation problems studied in [4, 5, 6], the Boolean inverse satisfiability problem [28], the characterisation of all polynomially closed Boolean relational clones [29], or the search for a surjective solution of a CSP. A possible approach to answer these questions or to even classify the complexity in such situations (many examples can be found in the previously listed references) is to consider weaker invariants that do not rely on existential quantification, e.g., strong partial clones, their counterparts—weak relational systems, and weak bases of relational clones [37, 38].

For each finitely generated Boolean relational clone, Victor Lagerkvist [24] determined a singleton weak base consisting of a certain 'minimal' relation. When treating computational problems related to Boolean clones via the weak bases approach, it often happens—exemplified in the investigations in [6], but also in the context of counting complexity when one restricts the allowable reductions—that one wishes to deal with groups of closely related Boolean clones, e.g., Horn clones $\mathsf{E}, \mathsf{E}_0, \mathsf{E}_1, \mathsf{E}_2$, cf. Figure 1, in a uniform way. Due to the minimality requirement on the number of tuples imposed in [24], the weak bases from [24] do not always allow such a treatment, and this is addressed in the present article: we provide a partially new collection of weak bases for Boolean relational clones, fixing some misprints of [24] on the way. We believe that this will make studying complexity questions via the weak bases approach simpler and possibly more elegant. Our choice of groups of Boolean clones that we wish to treat in a similar fashion is clearly a subjective one; however, it is informed by similarities of the associated relational clones, by shared algebraic properties of the corresponding clones, and by experience in dealing with previous complexity classifications. For the convenience of the reader we give an overview of the relations we use in Appendix A.

We conclude the paper by exhibiting several expressibility results between our weak base relations belonging to different Boolean relational clones. Here we exploit the uniform shape of the weak base relations for the chosen groups of Boolean clones (cf. Figure 1), and we are convinced that such relationships will be useful in proving reductions (e.g., reductions with constant or linear growth of the parameter measuring the complexity as used in [22, 25, 19, 20, 21] or suitable variants thereof) between computational problems parametrised by generating systems of the respective Boolean relational clones.

2 Notation and preliminaries

We write $\mathbb{N} = \{0, 1, 2, \ldots\}$ for the set of natural numbers and $\mathbb{N}_+$ for $\mathbb{N} \setminus \{0\}$. The cardinality of a set A is written as $|A|$. A *partial function* between sets A and B formally is a triple $f = (A, f^\bullet, B)$ where $f^\bullet \subseteq A \times B$ is a *right-unique* relation, that is, $f^\bullet$ satisfies $y = y'$ for all pairs $(x, y), (x, y') \in f^\bullet$. The *domain* of f is the set $\operatorname{dom}(f) := \{x \in A \mid \exists y \in B \colon (x, y) \in f^\bullet\}$. To simplify notation, we often write $f \colon D \subseteq A \longrightarrow B$ to denote that $f = (A, f^\bullet, B)$ is a partial function with domain D. A *(total) function* between A and B is a partial function f, where $\operatorname{dom}(f) = A$.

For an integer $n \in \mathbb{N}$, an n-ary *partial operation* on a set A is a partial function $f \colon D \subseteq A^n \longrightarrow A$, and an n-ary *operation* on A is any total function $f \colon A^n \longrightarrow A$. We collect the former in the set $\mathcal{P}_A^{(n)}$ and the latter in $\mathcal{O}_A^{(n)}$; moreover we set $\mathcal{O}_A := \bigcup_{n \in \mathbb{N}_+} \mathcal{O}_A^{(n)}$ and $\mathcal{P}_A := \bigcup_{n \in \mathbb{N}_+} \mathcal{P}_A^{(n)}$. For a set $F \subseteq \mathcal{P}_A$, we define $F^{(n)}$ to be $\mathcal{P}_A^{(n)} \cap F$. Also, if $f, g \in \mathcal{P}_A$, we say that f is a *subfunction* of g and write $f \preceq g$ if they share the same arity and $f^\bullet \subseteq g^\bullet$, that is, if f is the restriction of g to $\operatorname{dom}(f)$ being a subset of $\operatorname{dom}(g)$. For $1 \leq i \leq n \in \mathbb{N}_+$ the *i-th n-ary projection* on A is the operation $e_i^{(n)} \in \mathcal{O}_A^{(n)}$ given by the rule $e_i^{(n)}(x_1, \ldots, x_n) := x_i$ for all $(x_1, \ldots, x_n) \in A^n$. We define $\mathcal{J}_A := \left\{ e_i^{(n)} \,\middle|\, 1 \leq i \leq n \in \mathbb{N}_+ \right\}$ as the set of all total projections on A; any subfunction of a projection in $\mathcal{J}_A$ is called a *partial projection*. Furthermore, if $n, m \in \mathbb{N}$ and $f \in \mathcal{P}_A^{(n)}$, $g_1, \ldots, g_n \in \mathcal{P}_A^{(m)}$, we define their *composition* to be the partial operation $f \circ (g_1, \ldots, g_n) \colon D \subseteq A^m \longrightarrow A$ having domain $D = \{\boldsymbol{x} \in \bigcap_{i=1}^n \operatorname{dom}(g_i) \mid (g_1(\boldsymbol{x}), \ldots, g_n(\boldsymbol{x})) \in \operatorname{dom}(f)\}$ and operating according to $f \circ (g_1, \ldots, g_n)(\boldsymbol{x}) := f(g_1(\boldsymbol{x}), \ldots, g_n(\boldsymbol{x}))$ for all $\boldsymbol{x} \in D$. If $f, g_1, \ldots, g_n \in \mathcal{O}_A$, then clearly $f \circ (g_1, \ldots, g_n) \in \mathcal{O}_A$, as well.

A *clone of partial operations*, or *partial clone* for short, is any subset $P \subseteq \mathcal{P}_A$ with $\mathcal{J}_A \subseteq P$ that is closed under composition. A partial clone $P \subseteq \mathcal{P}_A$ is *strong* if it is additionally closed under taking subfunctions—equivalently, if it contains all partial projections. A *clone (of (total) operations)*, also *functional clone*, on A is any partial clone $F \subseteq \mathcal{O}_A$, that is, any sub-partial clone of the (partial) clone of all total operations $\mathcal{O}_A$. The set of all clones on A forms a closure system; as such it has an associated closure operator, which we denote by $F \mapsto \langle F \rangle_{\mathcal{O}_A}$ and which maps every set $F \subseteq \mathcal{O}_A$ to the least clone on A containing F.

The number of clones on the Boolean set $A = \{0,1\}$ is $\aleph_0$; these countably many clones were first described by Post [33] and form a lattice, which is depicted in Figure 1. By contrast, the number of clones on any finite set A of size $|A| \geq 3$ [18, 1], as well as the number of partial or even just strong partial clones on A with $|A| \geq 2$, is already $2^{\aleph_0}$ [14, Theorem 2].

For $m \in \mathbb{N}$, an *m-ary relation on A* is any subset $\rho \subseteq A^m$. For the powerset of A^m, i.e., the set of all such relations, we introduce the notation $\mathcal{R}_A^{(m)}$, and we define $\mathcal{R}_A := \bigcup_{m\in\mathbb{N}_+} \mathcal{R}_A^{(m)}$ as the set of all *finitary (non-nullary) relations on A*. We conveniently use a representation of $\rho = \{r_1,\ldots,r_n\} \in \mathcal{R}_A^{(m)}$ with at most $n \in \mathbb{N}_+$ tuples in A^m as an $(m \times n)$-matrix the columns of which are the tuples $r_1,\ldots,r_n \in \rho$. Such a representation is, of course, only unique up to permutation of columns (and possibly duplication of columns); certainly, if $n = |\rho|$, then no duplication is necessary. If $F \subseteq \mathcal{O}_A$ and $\rho \in \mathcal{R}_A^{(m)}$, then by $\Gamma_F(\rho) \subseteq A^m$ we denote the subuniverse of the algebra $\langle A; F\rangle^m$ generated by ρ, that is, the least F-invariant (v.i.) relation containing ρ. If $\rho = \{r_1,\ldots,r_n\}$ and F is a clone, then it is well known that $\Gamma_F(\rho) = \left\{ f \circ (r_1,\ldots,r_n) \;\middle|\; f \in F^{(n)} \right\}$, where for all $i \in \{1,\ldots,n\}$ we understand tuples $r_i\colon m \longrightarrow A$ as functions. The requirement that F be a clone is not a severe restriction since always $\Gamma_F(\rho) = \Gamma_{\langle F\rangle_{\mathcal{O}_A}}(\rho)$. This also means that for a functional clone $F \subseteq \mathcal{O}_A$, presented via some generating set $G \subseteq \mathcal{O}_A$ as $F = \langle G\rangle_{\mathcal{O}_A}$, we may write $\Gamma_F(\rho) = \Gamma_{\langle G\rangle_{\mathcal{O}_A}}(\rho) = \Gamma_G(\rho)$, and thus $\Gamma_F(\rho)$ can be computed by subpower closing $\rho \subseteq A^m$ under the operations given by G.

For finite A and $n \in \mathbb{N}$, the *n-th graphic of a clone F* is the relation $\Gamma_F(\chi_n)$ representing the n-ary part $F^{(n)}$ of the clone as follows: $\chi_n \subseteq A^q$, where $q := |A|^n$, is a $|A|^n$-ary relation given by the n (value tuples of the) n-ary projections; to enforce an unambiguous definition we represent χ_n as the $(q \times n)$-matrix whose rows are all tuples of A^n ordered lexicographically with respect to a fixed underlying linear order of A (for $A = \{0,\ldots,k-1\}$ we implicitly use the natural linear order of A here). More abstractly speaking, we fix a bijection $\beta\colon q = \{0,\ldots,q-1\} \longrightarrow A^n$ (e.g., the one given by lexicographic ordering); then we have $\chi_n = \left\{ e_i^{(n)} \circ \beta \;\middle|\; 1 \leq i \leq n \right\}$. The tuples in $\Gamma_F(\chi_n)$ are then exactly the value tuples of all functions in $F^{(n)}$, where the order of enumeration of the values is determined by the bijection β, that is to say, $\Gamma_F(\chi_n) = \left\{ f \circ (e_1^{(n)} \circ \beta,\ldots,e_n^{(n)} \circ \beta) = (f \circ (e_1^{(n)},\ldots,e_n^{(n)})) \circ \beta = f \circ \beta \;\middle|\; f \in F^{(n)} \right\}$.

For $m \in \mathbb{N}_+$ a relation $\rho \subseteq A^m$ is *primitive positively definable* from a set $Q \subseteq \mathcal{R}_A$ if it can be written as $\rho = \{ (x_1,\ldots,x_m) \in A^m \mid \exists y_1 \cdots \exists y_l\colon \boldsymbol{z}_1 \in \rho_1 \wedge \cdots \wedge \boldsymbol{z}_t \in \rho_t \}$ where $t,l \in \mathbb{N}$, $t \neq 0$, $\rho_1,\ldots,\rho_t \in Q \cup \{\Delta_A\}$ and $\boldsymbol{z}_1,\ldots,\boldsymbol{z}_t$ are variable tuples constructed from $x_1,\ldots,x_m,y_1,\ldots,y_l$. Here $\Delta_A := \{(x,x) \mid x \in A\}$ denotes the *equality relation* on A. A *quantifier-free primitive positive definition* is one where

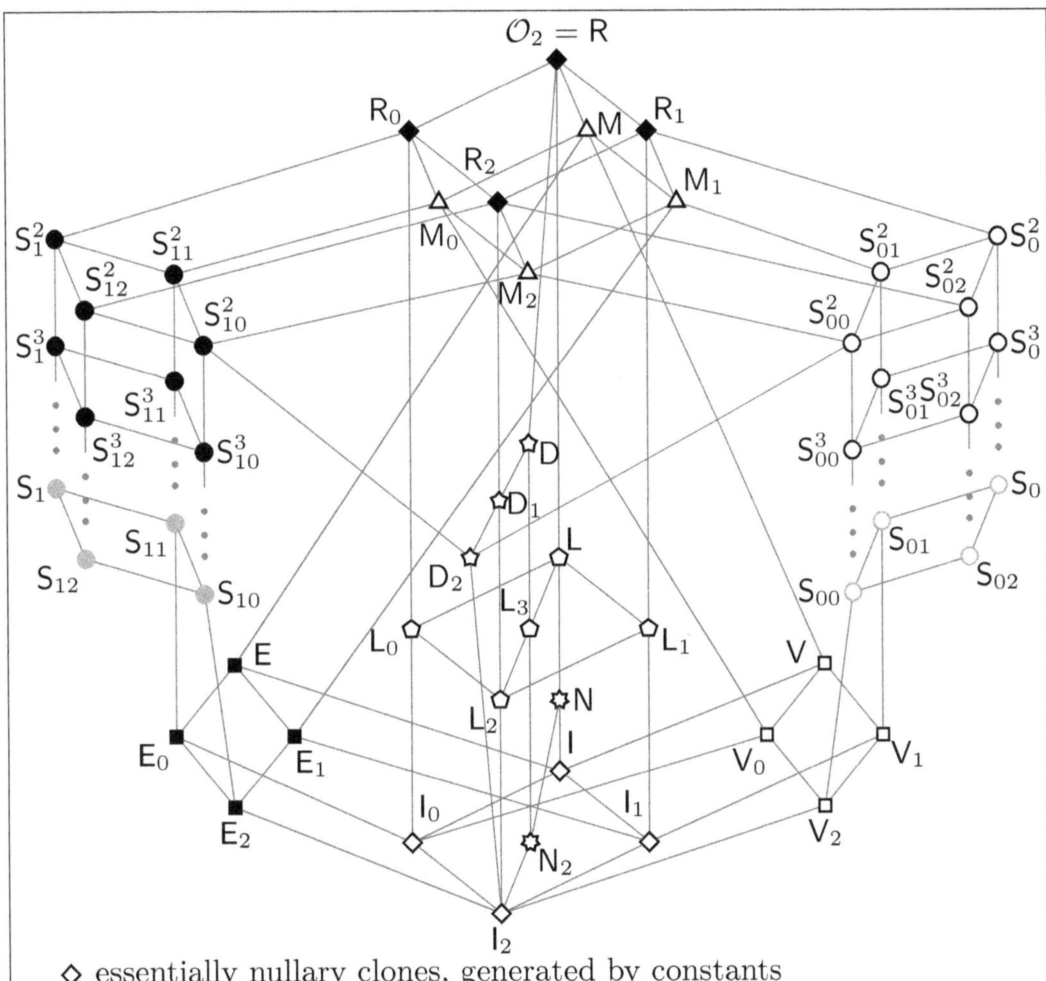

◇ essentially nullary clones, generated by constants
✡ essentially unary clones, containing negation
■ Horn clones, polynomially equivalent to a $\wedge$-semilattice
□ Dual Horn clones, polynomially equivalent to a $\vee$-semilattice
⬠ affine linear clones, polynomially equivalent to a vector space
☆ clones of selfdual, selfdual idempotent, selfdual monotone functions
△ clones of monotone operations
◆ clones of subset-preserving operations
○ clones of zero-separating functions (of various degrees)
● clones of one-separating functions (of various degrees)
●○ non-finitely related clones

Figure 1: Lattice of Boolean clones, 'similar' clones having identical nodes; naming conventions are guided by [8], where relational descriptions can be found

$l = 0$; it is *purely conjunctive* if, additionally, Δ_A is not used in it (unless it belongs to Q).

Example 1. We shall give here two examples how one calculates the n-th graphic of a clone (based on β given by the lexicographic ordering of n-tuples). We also show how calculations with the matrix representations of relations can be used to verify primitive positive expressions. These techniques are essential for many proofs of the paper; moreover, the expressions we derive here will be used later on.

The Boolean clone E is generated by the binary conjunction $\wedge$ and the two constant operations c_0, c_1 [15, Figure 2], that is, we have $\mathsf{E} = \langle\{\wedge, c_0, c_1\}\rangle_{\mathcal{O}_2}$. To compute the relation $\mathrm{R}_{\mathsf{E}} := \Gamma_{\mathsf{E}}(\chi_2)$ we start from the matrix representation of χ_2 and close the columns of the matrix componentwise under the generators of the clone E:

$$\chi_2 = \left\{\begin{matrix}00\\01\\10\\11\end{matrix}\right\} \rightsquigarrow \left\{\begin{matrix}0001\\0101\\1001\\1101\end{matrix}\right\} \rightsquigarrow \left\{\begin{matrix}00010\\01010\\10010\\11011\end{matrix}\right\} = \Gamma_{\mathsf{E}}(\chi_2).$$

The Boolean clone M consists of all monotone operations; the only unary operation on $\{0,1\}$ which is not monotone is the Boolean negation $\neg$. Hence we have $\mathsf{M}^{(1)} = \left\{e_1^{(1)}, c_0, c_1\right\}$ and thus $\mathrm{R}_{\mathsf{M}} := \Gamma_{\mathsf{M}}(\chi_1) = \left\{\begin{smallmatrix}001\\101\end{smallmatrix}\right\}$.

We now want to derive the relationship (cf. Lemma 31 below)

$$\forall x_1, x_2, x_3 \in \{0,1\}\colon \quad (x_1, x_2) \in \mathrm{R}_{\mathsf{M}} \iff (x_1, x_1, x_1, x_2) \in \mathrm{R}_{\mathsf{E}}\,.$$

For this we work with the matrix representation of R_{E} and suitably identify variables, i.e., rows of the matrix, deleting columns (marked by $\downarrow$) that violate the prescribed identification (and finally rearrange the rows according to the specified order of the variables and permute the columns such that the result matches the matrix derived above for R_{M}):

$$\begin{matrix}\overset{\downarrow\downarrow}{00}010 & x_1\\ 01010 & x_1\\ 10010 & x_1\\ 11011 & x_2\end{matrix} \rightsquigarrow \begin{matrix}010 & x_1\\ 010 & x_1\\ 010 & x_1\\ 011 & x_2\end{matrix} \rightsquigarrow \begin{matrix}010 & x_1\\ 011 & x_2\end{matrix} \rightsquigarrow \begin{matrix}001 & x_1\\ 101 & x_2\end{matrix}$$

In this example we verified a purely conjunctive expression, and thus we showed that $\mathrm{R}_{\mathsf{M}} \in [\{\mathrm{R}_{\mathsf{E}}\}]_{\wedge}$, see below for the meaning of this notation.

As a second example, we shall consider a formula that actually involves existential quantification. Namely, we want to demonstrate for any $x_1, \ldots, x_8 \in \{0,1\}$ that

$$\begin{aligned}&(x_1, \ldots, x_8) \in \Gamma_{\mathsf{L}_3}(\chi_3)\\ &\qquad\iff \exists! u, u', v, w \in \{0,1\}\colon (w, x_1, x_2, u, u', x_7, x_8, v) \in \Gamma_{\mathsf{L}_2}(\chi_3) \wedge\\ &\qquad\qquad\qquad (w, x_3, x_4, u, u', x_5, x_6, v) \in \Gamma_{\mathsf{L}_2}(\chi_3),\end{aligned}$$

cf. Lemma 28. The clone L_2 is generated by the ternary Mal'cev operation $g \in \mathcal{O}_2^{(3)}$ given by $g(x, y, z) := x \oplus y \oplus z$ (triple sum modulo 2) for all $x, y, z \in \{0, 1\}$, cf. [15, Figure 2]. Being the join of the clones L_2 and N_2 (cf. Figure 1), the clone L_3 is generated by $\{g, \neg\}$ since $\neg$ generates N_2, see [15, Figure 2]. We therefore obtain the octonary relations $\Gamma_{\mathsf{L}_2}(\chi_3)$ and $\Gamma_{\mathsf{L}_3}(\chi_3)$ by closing the three columns in χ_3 componentwise under g, and under g and $\neg$, respectively:

$$\chi_3 = \left\{\begin{matrix} 000\\ 001\\ 010\\ 011\\ 100\\ 101\\ 110\\ 111 \end{matrix}\right\} \rightsquigarrow \left\{\begin{matrix} 0000\\ 0011\\ 0101\\ 0110\\ 1001\\ 1010\\ 1100\\ 1111 \end{matrix}\right\} = \Gamma_{\mathsf{L}_2}(\chi_3), \quad \left\{\begin{matrix} 00001111\\ 00111100\\ 01011010\\ 01101001\\ 10010110\\ 10100101\\ 11000011\\ 11110000 \end{matrix}\right\} = \Gamma_{\mathsf{L}_3}(\chi_3)$$

In order to check the existentially quantified expression above, we form the sedenary relation $\Gamma_{\mathsf{L}_2}(\chi_3) \times \Gamma_{\mathsf{L}_2}(\chi_3)$, consisting of sixteen tuples, and identify variables as given in the expression above. We then remove the columns violating the variable identification (again marked by $\downarrow$) and observe for later use in Lemma 28 that the variables u, u', w, v always satisfy $u = x_1 \oplus x_2$, $u' = \neg u$, $w = 0$ and $v = 1$. We then project to the non-quantified variables, that is, we remove all rows labelled by existentially quantified variables, scilicet w, u, u', v. After that we reorder the rows according to ascending variable indices $x_1, \dots, x_8$, and finally, we reorder the columns in the matrix such that the resulting matrix exactly matches the representation computed above for $\Gamma_{\mathsf{L}_3}(\chi_3)$:

$$\begin{array}{ll} 0\overset{\downarrow}{0}\overset{\downarrow}{0}0\overset{\downarrow}{0}00\overset{\downarrow}{0}\overset{\downarrow}{0}00\overset{\downarrow}{0}0\overset{\downarrow}{0}\overset{\downarrow}{0}0 & w\\ 0000000011111111 & x_1\\ 0000111100001111 & x_2\\ 0000111111110000 & u\\ 1111000000001111 & u'\\ 1111000011110000 & x_7\\ 1111111100000000 & x_8\\ 1111111111111111 & v\\ 0000000000000000 & w\\ 0011001100110011 & x_3\\ 0101010101010101 & x_4\\ 0110011001100110 & u\\ 1001100110011001 & u'\\ 1010101010101010 & x_5\\ 1100110011001100 & x_6\\ 1111111111111111 & v \end{array} \rightsquigarrow \begin{array}{ll} 00000000 & w\\ 00001111 & x_1\\ 00110011 & x_2\\ 00111100 & u\\ 11000011 & u'\\ 11001100 & x_7\\ 11110000 & x_8\\ 11111111 & v\\ 00000000 & w\\ 01010101 & x_3\\ 01101001 & x_4\\ 00111100 & u\\ 11000011 & u'\\ 10010110 & x_5\\ 10101010 & x_6\\ 11111111 & v \end{array} \rightsquigarrow \begin{array}{ll} 00001111 & x_1\\ 00110011 & x_2\\ 11001100 & x_7\\ 11110000 & x_8\\ 01010101 & x_3\\ 01101001 & x_4\\ 10010110 & x_5\\ 10101010 & x_6 \end{array} \rightsquigarrow \begin{array}{ll} 00001111 & x_1\\ 00110011 & x_2\\ 01010101 & x_3\\ 01101001 & x_4\\ 10010110 & x_5\\ 10101010 & x_6\\ 11001100 & x_7\\ 11110000 & x_8 \end{array} \rightsquigarrow \begin{array}{ll} 00001111 & x_1\\ 00111100 & x_2\\ 01011010 & x_3\\ 01101001 & x_4\\ 10010110 & x_5\\ 10100101 & x_6\\ 11000011 & x_7\\ 11110000 & x_8 \end{array}$$

In many results of this article, we require the reader to perform routine calculations of this form on their own; in fact, this is what we mean if we write that a certain relationship follows from the matrix representations of the relations involved. Sometimes additional hints are given, but the matrix manipulations are usually omitted.

A *relational clone* on a finite set A is any subset $Q \subseteq \mathcal{R}_A$ that is closed under all relations that are primitive positively definable from it. Every relational clone Q on A contains Δ_A, and for $m, n \in \mathbb{N}_+$, if $\rho \subseteq A^m$ belongs to Q, then so do

$$V_\alpha(\rho) := \left\{ (x_{\alpha(1)}, \dots, x_{\alpha(n)}) \;\middle|\; (x_1, \dots, x_m) \in \rho \right\},$$
$$W_\beta(\rho) := \left\{ (x_1, \dots, x_n) \in A^n \;\middle|\; (x_{\beta(1)}, \dots, x_{\beta(m)}) \in \rho \right\}$$

for any maps $\alpha\colon \{1, \dots, n\} \to \{1, \dots, m\}$, $\beta\colon \{1, \dots, m\} \to \{1, \dots, n\}$; these relations are even definable from Q by a quantifier-free primitive positive formula for arbitrary β and for surjective α. If we just require closure of a set $Q \subseteq \mathcal{R}_A$ under quantifier-free primitive positive definitions, we obtain the more general notion of a *weak system with equality*; if we use only purely conjunctive (equality-free) definitions, we get *weak systems* of relations. It is easy to see that weak systems, and also relational clones, are moreover closed under finite intersection of relations of the same arity. The sets of all weak systems, and of all those with equality, and of all relational clones on A each form a closure system. For technical reasons we have to focus on weak systems / weak systems with equality / relational clones that also include the empty relation; we denote the corresponding closure operators by $Q \mapsto [Q]_\wedge$, $Q \mapsto [Q]_{\wedge,=}$, and $Q \mapsto [Q]_{\mathcal{R}_A}$, respectively, and they compute the least weak system / weak system with equality / relational clone containing $Q \cup \{\emptyset\}$.

There is an integral relationship between clones and relational clones, and between strong partial clones and weak systems with equality. It is described by the following Galois connection between partial operations and relations induced by the concept of *preservation*. For $m, n \in \mathbb{N}$ we say that $f \in \mathcal{P}_A^{(n)}$ *preserves* a relation $\rho \subseteq A^m$ if for every $(m \times n)$-matrix $X \in A^{m \times n}$ with columns $(r_1, \dots, r_n)$ and rows $(z_1, \dots, z_m)$ the following condition holds: if $r_1, \dots, r_n \in \rho$ and $z_1, \dots, z_m \in \operatorname{dom}(f)$, then also $f \circ (r_1, \dots, r_n) := (f(z_1), \dots, f(z_m)) \in \rho$. If this condition holds, we write $f \rhd \rho$. Clearly, if $f \in \mathcal{O}_A^{(n)}$ is total, the implication that needs to be checked is that $f \circ (r_1, \dots, r_n) \in \rho$ whenever $r_1, \dots, r_n \in \rho$. If $Q \subseteq \mathcal{R}_A$ is a set of relations, the (partial) operations preserving all relations in Q are called *(partial) polymorphisms* of Q; conversely, the preserved relations are called *invariants*. We use the following standard notation: $\operatorname{Pol} Q = \{ f \in \mathcal{O}_A \mid \forall \rho \in Q\colon f \rhd \rho \}$ $\operatorname{pPol} Q = \{ f \in \mathcal{P}_A \mid \forall \rho \in Q\colon f \rhd \rho \}$ for polymorphisms and partial polymorphisms of $Q \subseteq \mathcal{R}_A$, respectively, and $\operatorname{Inv} P = \{ \rho \in \mathcal{R}_A \mid \forall f \in P\colon f \rhd \rho \}$ for the invariant relations of a set of partial (or total) operations $P \subseteq \mathcal{P}_A$. We observe that $\mathcal{O}_A \cap \operatorname{pPol} Q = \operatorname{Pol} Q$.

For a finite set A, the following fundamental theorems describe the Galois closed sets of partial and total operations.

Theorem 2 ([7, 17]). *For a finite set A the clones $F \subseteq \mathcal{O}_A$ are exactly the Galois closed sets of the form $F = \operatorname{Pol}\operatorname{Inv} F$. Moreover, the Galois closed sets $Q = \operatorname{Inv}\operatorname{Pol} Q \subseteq \mathcal{R}_A$ are precisely all relational clones that include the empty relation.*

Theorem 3 ([17, 35, 34]). *For a finite set A, the strong partial clones $P \subseteq \mathcal{P}_A$ are exactly the Galois closed sets of the form $P = \operatorname{pPol}\operatorname{Inv} P$. Moreover, the Galois closed sets $Q = \operatorname{Inv}\operatorname{pPol} Q \subseteq \mathcal{R}_A$ are precisely all weak systems with equality, also containing the empty relation.*

Note that in both theorems the necessity to include the empty relation in the Galois closed sets on the relational side comes from the choice to *not* include nullary (partial) operations on the other side of the Galois correspondence.

In [38, 37] it is proved that for finite A the union of the set

$$\mathcal{L}_{\restriction F} := \{P \subseteq \mathcal{P}_A \mid P \text{ strong partial clone}, \mathcal{O}_A \cap P = F\}$$

(that is, $\bigcup \mathcal{L}_{\restriction F} = \bigcup\{\operatorname{pPol} R \mid R \subseteq \mathcal{R}_A, \operatorname{Pol} R = F\}$) over all strong partial clones that share the same clone $F \subseteq \mathcal{O}_A$ as their total part, is again a strong partial clone. It trivially follows from this that $\bigcup \mathcal{L}_{\restriction F} \in \mathcal{L}_{\restriction F}$; hence $\mathcal{L}_{\restriction F}$ has a largest element under inclusion. Under the Galois connections characterised in Theorems 2 and 3, the dual statement of this is that there is a least weak system with equality $S_F \subseteq \mathcal{R}_A$ having the property that $[S_F]_{\mathcal{R}_A} = \operatorname{Inv} F$, scilicet

$$S_F = \operatorname{Inv} \bigcup \mathcal{L}_{\restriction F} = \bigcap_{\substack{R \subseteq \mathcal{R}_A \\ \operatorname{Pol} R = F}} \operatorname{Inv}\operatorname{pPol} R = \bigcap_{\substack{R \subseteq \mathcal{R}_A \\ \operatorname{Pol} R = F}} [R]_{\wedge,=}.$$

Coming from a computational perspective, Schnoor and Schnoor [37, 38] define a *weak base* of $\operatorname{Inv} F$ as any *finite* set $W \subseteq \mathcal{R}_A$ of relations such that $S_F = [W]_{\wedge,=}$. Lagerkvist and Roy [27] have recently relaxed the finiteness requirement and call any $W \subseteq \mathcal{R}_A$ with $[W]_{\wedge,=} = S_F$ a weak base. We stick here to the original definition. Although it is conceptually misleading, for the sake of convenience, we agree, for a clone F and a set of relations W, that saying 'W is a weak base of F' means that W is a weak base of $\operatorname{Inv} F$. Note that every weak base W of $\operatorname{Inv} F$ is a generating system of $\operatorname{Inv} F$, for $\operatorname{Inv} F = [S_F]_{\mathcal{R}_A} = \left[[W]_{\wedge,=}\right]_{\mathcal{R}_A} = [W]_{\mathcal{R}_A}$.

In [37, Proposition 5.2] the relevance of irredundant relations was also noted: $\rho \subseteq A^m$ is *irredundant* if and only if for all $i, j \in m$ we have $i = j$ given that $a_i = a_j$ for all $(a_0, \ldots, a_{m-1}) \in \rho$. For $n \in \mathbb{N}_+$ and $q = |A|^n$, any q-ary relation $\rho \supseteq \chi_n$ is irredundant since for $i, j \in q$, $i \neq j$ the n-tuples $\beta(i), \beta(j)$ are distinct, so there is

$\ell \in \{1,\dots,n\}$ such that $e_\ell^{(n)} \circ \beta(i) \neq e_\ell^{(n)} \circ \beta(j)$ and $e_\ell^{(n)} \circ \beta \in \chi_n \subseteq \rho$. Moreover, for all $R \subseteq \mathcal{R}_A$, it follows from [37, Proposition 5.2] that any irredundant $\rho \subsetneq A^m$ belonging to $[R]_{\wedge,=}$ actually is in $[R]_{\wedge}$, that is, ρ can be expressed in terms of the relations from R without explicitly using the equality predicate. This is clearly very useful when dealing with problems that do not enjoy a priori compatibility with equality. We note that if we build a reduction for a CSP-like problem based on replacing in a formula atoms corresponding to a finite set W of relations by a conjunction of atoms from $R \cup \{=\}$ as consequence of an expressibility result such as $W \subseteq [R]_{\wedge,=}$ (or $W \subseteq [R]_{\wedge}$), the number of variables does not change and the growth in size of the formula is bounded above by a constant factor (namely, there is one of the finitely many relations in W having the largest growth factor comparing the length of atoms of the relation to their conjunctive representation, and this factor bounds the growth of the whole formula in the reduction).

Lemma 4. *For a finite set A and any set of relations $R \subseteq \mathcal{R}_A$, we have $W \subseteq [R]_{\wedge,=}$ for any weak base W of $[R]_{\mathcal{R}_A}$. If W contains only irredundant proper relations, we even obtain $W \subseteq [R]_{\wedge}$.*

Proof. Define $F := \operatorname{Pol} R$ and $S_F := \bigcap_{\substack{R' \subseteq \mathcal{R}_A \\ \operatorname{Pol} R' = F}} [R']_{\wedge,=}$; then, by the definition of weak base, we clearly have $W \subseteq [W]_{\wedge,=} = S_F \subseteq [R]_{\wedge,=}$. The additional statement about W consisting of irredundant relations $\rho \subsetneq A^m$ is easy and follows from [37, Proposition 5.2]. $\square$

Schnoor and Schnoor [37] define a *core* of a clone $F \subseteq \mathcal{O}_A$ as any relation $\rho \in \mathcal{R}_A$ such that $\operatorname{Pol}\{\Gamma_F(\rho)\} = F$, and $|\rho|$ as one of several possible *core sizes* of F. They suggest on p. 240 et seq. to treat *finitely related* clones $F = \operatorname{Pol} Q_0$, where $Q_0 = \{\rho_1,\dots,\rho_t\} \subseteq \mathcal{R}_A \setminus \{\emptyset\}$ with $t \in \mathbb{N}$ is a finite set of non-empty relations, as follows: letting $\rho := \rho_1 \times \cdots \times \rho_t$ we have $F = \operatorname{Pol}\{\rho\}$. Since ρ is hence F-invariant, we have $\Gamma_F(\rho) = \rho$ and thus $F = \operatorname{Pol}\{\rho\} = \operatorname{Pol}\{\Gamma_F(\rho)\}$, turning ρ into a core and giving $|\rho|$ as a core size of F. As core sizes are essential for the description of weak bases (cf. [37, Theorem 4.11]), and taking direct products is highly inefficient in this respect, we improve on this construction with the first result of the subsequent section.

3 Construction of singleton weak bases

Theorem 5. *Let A be a finite set, $n \in \mathbb{N}_+$, and $F := \operatorname{Pol} Q$ with $Q \subseteq \mathcal{R}_A$ be such that for each $\rho \in Q$ there is a set $B_\rho \subseteq \rho$ of size $1 \leq |B_\rho| \leq n$ satisfying $\rho = \Gamma_F(B_\rho)$. Then χ_n is a core of F of size n, i.e., $F = \operatorname{Pol}\{\Gamma_F(\chi_n)\}$.*

Proof. By its construction, we have $\Gamma_F(\chi_n) \in \operatorname{Inv} F$, thus $F \subseteq \operatorname{Pol}\{\Gamma_F(\chi_n)\}$. For the converse inclusion consider $\ell \in \mathbb{N}_+$ and $f \in \operatorname{Pol}\{\Gamma_F(\chi_n)\}$ of arity ℓ; we have to show that $f \in \operatorname{Pol} Q$. For this take $\rho = \Gamma_F(B_\rho) \in Q$ with $B_\rho = \{r_1, \dots, r_n\} \subseteq A^m$ and tuples $s_1, \dots, s_\ell \in \rho = \Gamma_F(B_\rho)$. Then there are functions $g_1, \dots, g_\ell \in F^{(n)}$ such that $s_j = g_j \circ (r_1, \dots, r_n)$ for each $j \in \{1, \dots, \ell\}$. We consider $r_1, \dots, r_n$ as columns of an $(m \times n)$-matrix with rows $z_1, \dots, z_m \in A^n$. Moreover, let $q := |A|^n$ and let $\beta\colon q \longrightarrow A^n$ be the bijection used in the definition of χ_n; define $\alpha(i) := \beta^{-1}(z_i)$ for $i \in \{1, \dots, m\}$. Hence, we have $e_j^{(n)}(z_i) = e_j^{(n)} \circ \beta \circ \beta^{-1}(z_i) = e_j^{(n)} \circ \beta(\alpha(i))$ for all $j \in \{1, \dots, n\}$ and all $i \in \{1, \dots, m\}$. As $e_j^{(n)}(z_i)$ is the i-th entry of r_j, we thus obtain $r_j = e_j^{(n)} \circ \beta \circ \alpha$ for all $j \in \{1, \dots, n\}$. Consequently,

$$s_t = g_t \circ (e_1^{(n)} \circ \beta \circ \alpha, \dots, e_n^{(n)} \circ \beta \circ \alpha) = g_t \circ (e_1^{(n)}, \dots, e_n^{(n)}) \circ \beta \circ \alpha = g_t \circ \beta \circ \alpha$$

for all $t \in \{1, \dots, \ell\}$. We have $g_1, \dots, g_\ell \in F \subseteq \operatorname{Pol}\{\Gamma_F(\chi_n)\} \ni f$, and therefore $h := f \circ (g_1, \dots, g_\ell) \in \operatorname{Pol}\{\Gamma_F(\chi_n)\}$. Hence $h \circ \beta = h \circ (e_1^{(n)} \circ \beta, \dots, e_n^{(n)} \circ \beta) \in \Gamma_F(\chi_n)$, that is, by the characterisation of $\Gamma_F(\chi_n)$, we infer $h \circ \beta = g \circ \beta$ for some $g \in F^{(n)}$. As β is bijective, it follows that $h = g \in F$. Now we can finally conclude

$$\begin{aligned} f \circ (s_1, \dots, s_\ell) &= f \circ (g_1 \circ (r_1, \dots, r_n), \dots, g_\ell \circ (r_1, \dots, r_n)) \\ &= f \circ (g_1, \dots, g_\ell) \circ (r_1, \dots, r_n) = h \circ (r_1, \dots, r_n) \in \Gamma_F(B_\rho) = \rho, \end{aligned}$$

where the final containment holds since $h \in F$. □

Note that in the situation described at the end of Section 2 we hence get $\max\{1, |\rho_1|, \dots, |\rho_t|\}$ instead of $|\rho_1| \cdot \ldots \cdot |\rho_t|$ as a core size of a finitely related $F = \operatorname{Pol}\{\rho_1, \dots, \rho_t\}$ with $t \geq 0$.

Corollary 6. *Let A be a finite set, $F := \operatorname{Pol} Q$ with $Q \subseteq \mathcal{R}_A$ be such that for each $\rho \in Q$ there is a set $B_\rho \subseteq \rho$ of size $1 \leq |B_\rho| \leq n \in \mathbb{N}_+$ satisfying $\rho = \Gamma_F(B_\rho)$. Then $\{\Gamma_F(\chi_n)\}$ is a weak base of F consisting of a single irredundant relation.*

It is possible to appeal to [37, Theorem 4.11] to infer that $\Gamma_F(\chi_n)$ provides a singleton weak base from the fact that F has core size n (see Theorem 5). However, using the theory developed above, we can also give a simple self-contained argument:

Proof. We use Theorem 5 to get that $F = \operatorname{Pol}\{\Gamma_F(\chi_n)\}$. We will show that $\operatorname{pPol}\{\Gamma_F(\chi_n)\} = \bigcup \mathcal{L}_{\restriction F} = \bigcup_{\substack{R \subseteq \mathcal{R}_A \\ \operatorname{Pol} R = F}} \operatorname{pPol} R$; thus $[\{\Gamma_F(\chi_n)\}]_{\wedge,=} = \operatorname{Inv} \operatorname{pPol}\{\Gamma_F(\chi_n)\}$ equals the least weak system S_F with $[S_F]_{\mathcal{R}_A} = \operatorname{Inv} F$. Due to $F = \operatorname{Pol}\{\Gamma_F(\chi_n)\}$, the inclusion $\operatorname{pPol}\{\Gamma_F(\chi_n)\} \subseteq \bigcup \mathcal{L}_{\restriction F}$ is obvious. For the converse, let $\ell \in \mathbb{N}_+$ and suppose an ℓ-ary f satisfies $f \in \operatorname{pPol} R$ for some $R \subseteq \mathcal{R}_A$ with $\operatorname{Pol} R = F$. To show that $f \in \operatorname{pPol}\{\Gamma_F(\chi_n)\}$, consider $g_1, \ldots, g_\ell \in F^{(n)}$ so that $g_1 \circ \beta, \ldots, g_\ell \circ \beta \in \Gamma_F(\chi_n)$ and $(g_1(\beta(i)), \ldots, g_\ell(\beta(i))) \in \operatorname{dom}(f)$ for every $i \in q = |A|^n$. Since β is bijective, this means that $(g_1(z), \ldots, g_\ell(z)) \in \operatorname{dom}(f)$ for all $z \in A^n$; therefore, the composition $h := f \circ (g_1, \ldots, g_\ell) \in \mathcal{O}_A$ is total. Since $R \subseteq \mathcal{R}_A$ satisfies $F = \operatorname{Pol} R$, we have $g_1, \ldots, g_\ell \in F = \operatorname{Pol} R \subseteq \operatorname{pPol} R$, but also $f \in \operatorname{pPol} R$ due to the assumption on f. Hence, we get $h = f \circ (g_1, \ldots, g_\ell) \in \mathcal{O}_A \cap \operatorname{pPol} R = \operatorname{Pol} R = F$, wherefore $f \circ (g_1 \circ \beta, \ldots, g_\ell \circ \beta) = f \circ (g_1, \ldots, g_\ell) \circ \beta = h \circ \beta \in \Gamma_F(\chi_n)$ as desired.

This proves that $\Gamma_F(\chi_n)$ constitutes a weak base of $\operatorname{Inv} F$; it is an irredundant relation because $\chi_n \subseteq \Gamma_F(\chi_n)$. □

To construct new weak bases from old ones, we use the following lemma.

Lemma 7. *If $W \subseteq \mathcal{R}_A$ is a weak base of a clone $F \subseteq \mathcal{O}_A$ on a finite set A and some finite $W' \subseteq [W]_{\wedge,=}$ is such that* $\operatorname{Pol} W' = F$, *then W' is a weak base of F, too.*

Note that under the assumptions of the lemma we have $W' \subseteq [W]_{\mathcal{R}_A}$, implying by Theorem 2 that $F = \operatorname{Pol}\operatorname{Inv} F = \operatorname{Pol}[W]_{\mathcal{R}_A} = \operatorname{Pol} W \subseteq \operatorname{Pol} W'$ because the weak base W generates the relational clone $\operatorname{Inv} F$. Thus, the precise condition that needs to be verified when applying Lemma 7 is that this inclusion is not proper, i.e., that $\operatorname{Pol} W' \subseteq F$.

Proof. We abbreviate $Q := \operatorname{Inv} F$. As $\operatorname{Pol} W' = F$, we get $[W']_{\mathcal{R}_A} = \operatorname{Inv}\operatorname{Pol} W' = Q$; therefore, $W \subseteq [W']_{\wedge,=}$ by the weak base property w.r.t. Q (cf. Lemma 4). This implies $[W]_{\wedge,=} \subseteq [W']_{\wedge,=}$, and together with the assumption $[W']_{\wedge,=} \subseteq [W]_{\wedge,=}$ we obtain $[W']_{\wedge,=} = [W]_{\wedge,=} = S_F$. Therefore, W' also is a weak base of F. □

Corollary 8. *If $W \subseteq \mathcal{R}_A$ is a weak base of a clone $F \subseteq \mathcal{O}_A$ and $W' \subseteq \mathcal{R}_A$ arises from W by subjecting each $\rho \in W$ to some permutation of its entries, then W' is a weak base of F, as well.*

4 Boolean weak bases with similarities

We denote the clones of Post's lattice by the symbols shown in Figure 1. Often we write $2 = \{0, 1\}$ for the Boolean set; for example, $\mathcal{L}_2 = \mathcal{L}_{\{0,1\}}$ means the set of all Boolean clones. We further use $\wedge, \vee$ for the binary Boolean conjunction and disjunction, $\oplus$ for addition modulo 2 (exclusive disjunction), $\rightarrow$ for implication, $\neg, c_0, c_1$ for negation and the two unary constant operations, and $\boldsymbol{a} = (a, \ldots, a) \in 2^n$ where $a \in \{0, 1\}$ for the two constant tuples, their arity n usually being implicit.

If $F \in \mathcal{L}_2$, then we write R_F for the relation given in [24, Table 1] such that $\{\mathrm{R}_F\}$ is a weak base of F, and ρ_F in the cases where we give an alternative weak base $\{\rho_F\}$. For most relations we use in this paper, descriptions as sets of tuples are given in Appendix A. Clones $F \in \mathcal{L}_2$ that we wish to treat as a 'similar' group are those with a closely related generating set, or with a similar generating set for $\operatorname{Inv} F$. The groups we choose are clones generated by constants $(\mathsf{I}, \mathsf{I}_0, \mathsf{I}_1, \mathsf{I}_2)$, essentially unary clones involving negation $(\mathsf{N}, \mathsf{N}_2)$, Horn $(\mathsf{E}, \mathsf{E}_0, \mathsf{E}_1, \mathsf{E}_2)$ and dual Horn clones $(\mathsf{V}, \mathsf{V}_0, \mathsf{V}_1, \mathsf{V}_2)$, affine (linear) clones $(\mathsf{L}, \mathsf{L}_0, \mathsf{L}_1, \mathsf{L}_2, \mathsf{L}_3)$, clones between selfdual and selfdual monotone operations $(\mathsf{D}, \mathsf{D}_1, \mathsf{D}_2)$, clones of zero-separating functions $(\mathsf{S}_0^n, \mathsf{S}_{01}^n, \mathsf{S}_{02}^n, \mathsf{S}_{00}^n)$ and of one-separating functions $(\mathsf{S}_1^n, \mathsf{S}_{11}^n, \mathsf{S}_{12}^n, \mathsf{S}_{10}^n)$, monotone clones $(\mathsf{M}, \mathsf{M}_0, \mathsf{M}_1, \mathsf{M}_2)$, and subset-preserving clones $(\mathcal{O}_2 = \mathsf{R}, \mathsf{R}_0, \mathsf{R}_1, \mathsf{R}_2)$. In several cases, these clones come in groups of four that are ordered in the form of a diamond. In these cases we would preferably try to maintain constructibility of our weak base relations according to the commutative diagram specified by the upwards directed edges of the order of the corresponding relational clones, where parallel edges represent a common type of construction (ideally preserving the cardinality of the relation), cf. Figure 2. Moreover, if for some $F \in \mathcal{L}_2$ the weak base relation R_F is an argument permutation

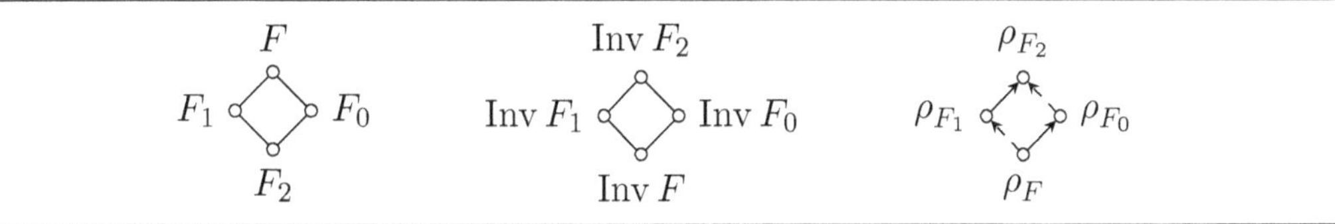

Figure 2: Square covering sublattices of Boolean relational clones and intended constructions of weak base relations

of $\Gamma_F(\chi_n)$ for some $n \in \mathbb{N}_+$, then we shall actually prefer to choose $\rho_F = \Gamma_F(\chi_n)$ with the argument order determined by the lexicographic order defining χ_n.

We start from the top of Post's lattice. For the subset-preserving clones, we wish to modify $\mathrm{R}_{\mathsf{R}} = \{(x, x) \mid x \in 2\}$, which is not irredundant. Our choice is irredundant, but contains a fictitious argument, allowing for a uniform construction.

Lemma 9. *We have the weak base relations* $\rho_{\mathsf{R}} = \{0, 1\}$, $\mathrm{R}_{\mathsf{R}_a} = \{a\}$ *for* $a \in 2$ *and* $\mathrm{R}_{\mathsf{R}_2} = \{(0, 1)\}$*; alternatively, the relations* ρ_{R}, $\rho_{\mathsf{R}_0} = \Gamma_{\mathsf{R}_0}(\chi_1) = \{0\} \times \rho_{\mathsf{R}}$, $\rho_{\mathsf{R}_1} = \Gamma_{\mathsf{R}_1}(\chi_1) = \rho_{\mathsf{R}} \times \{1\}$ *and* $\rho_{\mathsf{R}_2} = \{0\} \times \rho_{\mathsf{R}} \times \{1\} = \{0\} \times \rho_{\mathsf{R}_1} = \rho_{\mathsf{R}_0} \times \{1\}$ *also give weak bases.*

Proof. We have $\rho_{\mathsf{R}} = \{x \in 2 \mid (x, x) \in \mathrm{R}_{\mathsf{R}}\} \in [\{\mathrm{R}_{\mathsf{R}}\}]_{\wedge,=}$ and $\operatorname{Pol} \rho_{\mathsf{R}} = \mathcal{O}_2 = \mathsf{R}$, so ρ_{R} is a weak base by Lemma 7. By [24, Table 1] the four considered clones have core size 1, so we can apply [37, Theorem 4.11] or Corollary 6 to get the alternative irredundant weak bases for R_a, $a \in 2$. Taking $\mathrm{R}_{\mathsf{R}_2} = \{(0,1)\}$ from [24, Table 1], we can express the relation $\rho_{\mathsf{R}_2} = \{(x,y,z) \in 2^3 \mid (x,z) \in \mathrm{R}_{\mathsf{R}_2}\} \in [\{\mathrm{R}_{\mathsf{R}_2}\}]_{\wedge,=}$, and clearly we have $\mathrm{Pol}\,\{\rho_{\mathsf{R}_2}\} = \mathrm{Pol}\,\{\{0\},\{1\}\} = \mathsf{R}_0 \cap \mathsf{R}_1 = \mathsf{R}_2$; hence, $\{0\} \times \rho_{\mathsf{R}} \times \{1\}$ is a weak base of R_2 by Lemma 7. □

Lemma 10. *A weak base is given by the order relation* $\mathrm{R}_{\mathsf{M}} = \Gamma_{\mathsf{M}}(\chi_1) = \leq_2$, *and, similar to Lemma 9, it can be extended to* $\rho_{\mathsf{M}_0} = \{0\} \times \mathrm{R}_{\mathsf{M}}$, $\mathrm{R}_{\mathsf{M}_1} = \mathrm{R}_{\mathsf{M}} \times \{1\}$ *and* $\rho_{\mathsf{M}_2} = \{0\} \times \mathrm{R}_{\mathsf{M}} \times \{1\} = \{0\} \times \mathrm{R}_{\mathsf{M}_1} = \rho_{\mathsf{M}_0} \times \{1\}$, *describing singleton weak bases of the remaining clones from the monotone group.*

Proof. Clearly, we can express the relations $\rho_{\mathsf{M}_0} = \{\,(x,y,z) \in 2^3 \mid (y,z,x) \in \mathrm{R}_{\mathsf{M}_0}\}$, and $\rho_{\mathsf{M}_2} = \{\,(x,y,z,u) \in 2^3 \mid (y,z,x,u) \in \mathrm{R}_{\mathsf{M}_2}\}$ simply using variable permutations, wherefore the claim follows from Corollary 8. □

Lemma 11. *Weak bases are given by* $\mathrm{R}_{\mathsf{E}} = \Gamma_{\mathsf{E}}(\chi_2)$, *and, similar to Lemma 9, by* $\rho_{\mathsf{E}_0} = \{0\} \times \mathrm{R}_{\mathsf{E}}$, $\rho_{\mathsf{E}_1} = \mathrm{R}_{\mathsf{E}} \times \{1\}$ *and* $\rho_{\mathsf{E}_2} = \{0\} \times \mathrm{R}_{\mathsf{E}} \times \{1\} = \{0\} \times \rho_{\mathsf{E}_1} = \rho_{\mathsf{E}_0} \times \{1\}$.

Proof. Immediate verification gives that R_{E} from [24] equals $\Gamma_{\mathsf{E}}(\chi_2)$; moreover ρ_{E_0} provides a weak base due to $\rho_{\mathsf{E}_0} = \{\,(x,y,z,u,v) \in 2^5 \mid (y,z,u,v,x) \in \mathrm{R}_{\mathsf{E}_0}\}$ and Corollary 8. By Corollary 6, $\Gamma_{\mathsf{E}_1}(\chi_3)$ is a weak base relation for E_1 since the latter has core size $2 \leq 3$ (cf. [24, Table 1]). From $\Gamma_{\mathsf{E}_1}(\chi_3)$ we form

$$\{(x_1,\dots,x_5) \in 2^5 \mid (x_1,x_2,x_1,x_2,x_3,x_4,x_3,x_5) \in \Gamma_{\mathsf{E}_1}(\chi_3)\}$$

and direct verification shows that this relation is $\mathrm{R}_{\mathsf{E}} \times \{1\}$. We therefore know that $\rho_{\mathsf{E}_1} = \mathrm{R}_{\mathsf{E}} \times \{1\} \in [\{\Gamma_{\mathsf{E}_1}(\chi_3)\}]_{\wedge,=}$, thus $\rho_{\mathsf{E}_1} \in [\{\Gamma_{\mathsf{E}_1}(\chi_3)\}]_{\mathcal{R}_A} \subseteq [\mathrm{Inv}\,\mathsf{E}_1]_{\mathcal{R}_A} = \mathrm{Inv}\,\mathsf{E}_1$, which means that $\mathsf{E}_1 \subseteq \mathrm{Pol}\,\{\rho_{\mathsf{E}_1}\}$. This inclusion cannot be proper, because we have $\vee \notin \mathrm{Pol}\,\{\rho_{\mathsf{E}_1}\}$ (namely $(0,0,1,1,1) \vee (0,1,0,1,1) = (0,1,1,1,1) \notin \rho_{\mathsf{E}_1}$), showing $\mathsf{M}_1 \nsubseteq \mathrm{Pol}\,\{\rho_{\mathsf{E}_1}\}$, and moreover $(0,0,0,0,0) \notin \rho_{\mathsf{E}_1}$, i.e., $c_0 \notin \mathrm{Pol}\,\{\rho_{\mathsf{E}_1}\}$, showing $\mathsf{E} \nsubseteq \mathrm{Pol}\,\{\rho_{\mathsf{E}_1}\}$. As M_1 and E are the only upper covers of E_1 in Post's lattice, we infer $\mathrm{Pol}\,\{\rho_{\mathsf{E}_1}\} = \mathsf{E}_1$, and Lemma 7 proves that ρ_{E_1} is a weak base relation of E_1.

The proof for ρ_{E_2} is similar: $\Gamma_{\mathsf{E}_2}(\chi_4)$ is a weak base relation of E_2 for this clone has core size $3 \leq 4$ (cf. [24, Table 1]). We verify that $\rho_{\mathsf{E}_2} = \rho_{\mathsf{E}_0} \times \{1\} \in [\{\Gamma_{\mathsf{E}_2}(\chi_4)\}]_{\wedge,=}$, by checking that

$$\rho_{\mathsf{E}_2} = \Big\{(x_1,\dots,x_6) \in 2^6 \;\Big|\; (x_1,x_1,x_1,x_1,x_2,x_3,x_2,x_3,x_1,x_1,x_1,x_1,x_4,x_5,x_4,x_6) \in \Gamma_{\mathsf{E}_2}(\chi_4)\Big\}.$$

Thus $\rho_{\mathsf{E}_2} \in [\{\Gamma_{\mathsf{E}_2}(\chi_4)\}]_{\mathcal{R}_A} \subseteq [\mathrm{Inv}\,\mathsf{E}_2]_{\mathcal{R}_A} = \mathrm{Inv}\,\mathsf{E}_2$, i.e., $\mathsf{E}_2 \subseteq \mathrm{Pol}\,\{\rho_{\mathsf{E}_2}\}$. Certainly, $c_a \notin \mathrm{Pol}\,\{\rho_{\mathsf{E}_2}\}$ for both $a \in 2$, thus $\mathsf{E}_a \nsubseteq \mathrm{Pol}\,\{\rho_{\mathsf{E}_2}\}$. Neither is $(x,y,z) \mapsto x \wedge (y \vee z)$

in Pol $\{\rho_{\mathsf{E}_2}\}$ because

$$(0,1,1,1,1,1) \wedge ((0,0,0,1,1,1) \vee (0,0,1,0,1,1)) = (0,1,1,1,1,1) \wedge (0,0,1,1,1,1) = (0,0,1,1,1,1) \notin \rho_{\mathsf{E}_2}.$$

Therefore, we have $\mathsf{S}_{10} \nsubseteq \mathrm{Pol}\{\rho_{\mathsf{E}_2}\}$, cf. [15, Figure 2]. By inspecting Post's lattice, the inclusion $\mathsf{E}_2 \subseteq \mathrm{Pol}\{\rho_{\mathsf{E}_2}\}$ cannot be proper. Hence Lemma 7 shows that ρ_{E_2} is a weak base relation of E_2. □

Lemma 12. *Weak bases are given by $\rho_{\mathsf{V}} = \Gamma_{\mathsf{V}}(\chi_2)$,*[1] *and, similar to Lemma 9, by $\rho_{\mathsf{V}_1} = \{1\} \times \rho_{\mathsf{V}}$, $\rho_{\mathsf{V}_0} = \rho_{\mathsf{V}} \times \{0\}$ and $\rho_{\mathsf{V}_2} = \{1\} \times \rho_{\mathsf{V}} \times \{0\} = \{1\} \times \rho_{\mathsf{V}_0} = \rho_{\mathsf{V}_1} \times \{0\}$.*

Proof. This follows from Lemma 11 by duality, i.e., switching the roles of 0 and 1. □

For the four clones generated by constants, we cannot do much. We have $\Gamma_{\mathsf{I}_2}(\chi_n) = \chi_n$, $\Gamma_{\mathsf{I}_a}(\chi_n) = \chi_n \cup \{\boldsymbol{a}\}$ for $a \in 2$, and $\Gamma_{\mathsf{I}}(\chi_n) = \chi_n \cup \{\mathbf{0},\mathbf{1}\}$ for all $n \in \mathbb{N}_+$. The weak bases from [24] satisfy for all $x_1, x_2, x_3, x_4 \in 2$ that $(x_1,x_2,x_3,x_4) \in \mathrm{R}_{\mathsf{I}_1}$ iff there are (unique) $x_1', x_2', x_3', x_4' \in 2$ with $(x_1', x_2', x_3, x_1, x_2, x_3', x_4', x_4) \in \mathrm{R}_{\mathsf{I}_2}$ and $x_4' = 0$; also $(x_1,x_2,x_3,x_4) \in \mathrm{R}_{\mathsf{I}_0}$ iff there are (unique) $x_1', x_2', x_3', x_4' \in 2$ with $(x_3', x_2, x_1, x_3, x_2', x_1', x_4, x_4') \in \mathrm{R}_{\mathsf{I}_2}$ and $x_4' = 1$. This is to say that for $a \in \{0,1\}$ one can define $\mathrm{R}_{\mathsf{I}_a}$ from $\mathrm{R}_{\mathsf{I}_2}$ by existentially quantifying four variables, and that in each case the values of these four variables are in fact uniquely determined because both relations contain precisely three tuples. It therefore makes no difference whether one employs the quantifier $\exists$ or $\exists!$ in the definition; hence, when counting solutions plays a role, such a relationship can be exploited to get a parsimonious reduction. Choosing $\rho_{\mathsf{I}_a} = \chi_3 \cup \{\boldsymbol{a}\}$ for both $a \in 2$ and $\rho_{\mathsf{I}_2} = \chi_4$, we can achieve a similar property also for all four clones $\mathsf{I}, \mathsf{I}_0, \mathsf{I}_1, \mathsf{I}_2$.

Lemma 13. *For all $x_1, \dots, x_8 \in 2$ we have $(x_1,x_2,x_3,x_4) \in \mathrm{R}_{\mathsf{I}} = \Gamma_{\mathsf{I}}(\chi_2)$ iff there are (unique) values $x_1', x_2', x_3', x_4' \in 2$ such that $(x_1', x_2', x_3', x_4', x_1, x_2, x_3, x_4) \in \rho_{\mathsf{I}_0}$ (and $x_1' = 0$), equivalently, iff there are (unique) $x_1', x_2', x_3', x_4' \in 2$ such that ($x_4' = 1$ and) $(x_1, x_2, x_3, x_4, x_1', x_2', x_3', x_4') \in \rho_{\mathsf{I}_1}$. Moreover, we have $(x_1, x_2, \dots, x_8) \in \rho_{\mathsf{I}_0}$ iff there are (unique) values $x_1', \dots, x_8' \in 2$, such that ($x_i' = \neg x_i$ for $1 \leq i \leq 8$ and) $(x_1, x_8', x_2, x_7', x_3, x_6', x_4, x_5', x_5, x_4', x_6, x_3', x_7, x_2', x_8, x_1')$ is in ρ_{I_2}. Dually, we have $(x_1, x_2, \dots, x_8) \in \rho_{\mathsf{I}_1}$ iff there are (unique) values $x_1', \dots, x_8' \in 2$, with ($x_i' = \neg x_i$ for $i \leq 8$ and) $(x_8', x_1, x_7', x_2, x_6', x_3, x_5', x_4, x_4', x_5, x_3', x_6, x_2', x_7, x_1', x_8) \in \rho_{\mathsf{I}_2}$.*

Proof. This can be directly verified by writing out the matrix representations of the relations. The uniqueness of the values of the existentially quantified variables follows from the fact that all compared relations contain exactly four tuples. □

[1] In contrast to $\mathrm{R}_{\mathsf{E}} = \Gamma_{\mathsf{E}}(\chi_2)$, the relation R_{V} given by Lagerkvist in [24, Table 1] is only equal to $\Gamma_{\mathsf{V}}(\chi_2)$ up to a permutation of variables. Therefore in this lemma the notation ρ_{V} was used.

The weak base relations for the essentially unary clones with negation both have six tuples and are closely related, too.

Lemma 14. *The weak base relations for* N *and* N_2 *satisfy for all* $x_1, x_2, x_3, x_4 \in 2$ *that* $(x_1, x_2, x_3, x_4) \in \mathrm{R}_{\mathsf{N}} = \Gamma_{\mathsf{N}}(\chi_2)$ *iff there are unique elements* $x'_i = \neg x_i \in 2$ *(for* $1 \leq i \leq 4$*) such that* $(x_1, x_2, x_3, x_4, x'_1, x'_2, x'_3, x'_4) \in \mathrm{R}_{\mathsf{N}_2}$*, or alternatively such that* $(x_1, x_2, x_3, x_4, x'_4, x'_3, x'_2, x'_1) \in \rho_{\mathsf{N}_2} := \Gamma_{\mathsf{N}_2}(\chi_3)$.

Proof. This holds directly by the definition of $\mathrm{R}_{\mathsf{N}_2}$ in [24], which can be stated as

$$\mathrm{R}_{\mathsf{N}_2} = \left\{ (x_1, x_2, x_3, x_4, x'_1, x'_2, x'_3, x'_4) \in 2^8 \;\middle|\; (x_1, x_2, x_3, x_4) \in \mathrm{R}_{\mathsf{N}} \wedge \bigwedge_{i=1}^{4} x'_i = \neg x_i \right\}.$$

Moreover, for arbitrary $x_1, x_2, x_3, x_4, x'_1, x'_2, x'_3, x'_4 \in \{0,1\}$ it is easy to see that $(x_1, x_2, x_3, x_4, x'_1, x'_2, x'_3, x'_4) \in \mathrm{R}_{\mathsf{N}_2}$ iff $(x_1, x_2, x_3, x_4, x'_4, x'_3, x'_2, x'_1) \in \rho_{\mathsf{N}_2}$. □

For the clones D, D_1, D_2 we do not see an easy possibility to get weak base relations with the same number of tuples for all three of them, which might improve the relations from [24]. Thus, we stick with $\mathrm{R}_{\mathsf{D}} = \Gamma_{\mathsf{D}}(\chi_1)$, $\rho_{\mathsf{D}_1} = \Gamma_{\mathsf{D}_1}(\chi_2) = \chi_2$, being $\{ (x, y, z, u) \in 2^4 \mid (y, z, x, u) \in \mathrm{R}_{\mathsf{D}_1} \}$, and $\mathrm{R}_{\mathsf{D}_2}$; furthermore, since D_2 has core size 3 (cf. [24, Table 1]), we note that sometimes also $\rho_{\mathsf{D}_2} = \Gamma_{\mathsf{D}_2}(\chi_3)$ may be a useful weak base relation (see Corollary 6).

Lemma 15. *For* $n \in \mathbb{N}$, $n \geq 2$, *we have weak base relations* $\mathrm{R}_{\mathsf{S}_0^n} = (2^n \setminus \{\mathbf{0}\}) \times \{1\}$, $\rho_{\mathsf{S}_{02}^n} = \{0\} \times \mathrm{R}_{\mathsf{S}_0^n} = \{0\} \times (2^n \setminus \{\mathbf{0}\}) \times \{1\}$, $\mathrm{R}_{\mathsf{S}_{01}^n} = \rho_{\mathsf{S}_{02}^n} \cup \{\mathbf{1}\}$, $\rho_{\mathsf{S}_{00}^n} = \{0\} \times \mathrm{R}_{\mathsf{S}_{01}^n}$. *Dually, we have* $\mathrm{R}_{\mathsf{S}_1^n} = (2^n \setminus \{\mathbf{1}\}) \times \{0\}$, $\rho_{\mathsf{S}_{12}^n} = \{1\} \times \mathrm{R}_{\mathsf{S}_1^n} = \{1\} \times (2^n \setminus \{\mathbf{1}\}) \times \{0\}$, $\mathrm{R}_{\mathsf{S}_{11}^n} = \rho_{\mathsf{S}_{12}^n} \cup \{\mathbf{0}\}$, $\rho_{\mathsf{S}_{10}^n} = \{1\} \times \mathrm{R}_{\mathsf{S}_{11}^n}$.[2]

Proof. By Corollary 8, the relations $\rho_{\mathsf{S}_{02}^n}$ and $\rho_{\mathsf{S}_{00}^n}$ form weak bases because

$$\rho_{\mathsf{S}_{02}^n} = \left\{ (u, x_1, \ldots, x_n, v) \in 2^{n+2} \;\middle|\; (x_1, \ldots, x_n, u, v) \in \mathrm{R}_{\mathsf{S}_{02}^n} \right\},$$
$$\rho_{\mathsf{S}_{00}^n} = \left\{ (u, x_0, \ldots, x_n, v) \in 2^{n+3} \;\middle|\; (x_0, \ldots, x_n, u, v) \in \mathrm{R}_{\mathsf{S}_{00}^n} \right\}.$$

Following the definition of $\mathrm{R}_{\mathsf{S}_{01}^n}$ and $\mathrm{R}_{\mathsf{S}_{11}^n}$ from [24], we conclude

$$\mathrm{R}_{\mathsf{S}_{01}^n} = ((\{0\} \times (2^n \setminus \{\mathbf{0}\})) \cup \{\mathbf{1}\}) \times \{1\} = (\{0\} \times (2^n \setminus \{\mathbf{0}\}) \times \{1\}) \cup \{\mathbf{1}\}$$
$$= \rho_{\mathsf{S}_{02}^n} \cup \{\mathbf{1}\},$$

[2] Using $\rho_{\mathsf{S}_{12}^n} \cup \{\mathbf{0}\}$ as $\mathrm{R}_{\mathsf{S}_{11}^n}$ and $\rho_{\mathsf{S}_{10}^n}$ (or also $(\rho_{\mathsf{S}_{12}^n} \cup \{\mathbf{0}\}) \times \{1\}$) fixes the base relations for S_{11}^n and S_{10}^n that are incorrectly printed in [24, Table 1].

$$\mathrm{R}_{\mathsf{S}_{11}^n} = ((\{1\} \times (2^n \setminus \{\mathbf{1}\})) \cup \{\mathbf{0}\}) \times \{0\} = (\{1\} \times (2^n \setminus \{\mathbf{1}\}) \times \{0\}) \cup \{\mathbf{0}\}$$
$$= \rho_{\mathsf{S}_{12}^n} \cup \{\mathbf{0}\}.$$

The weak base relations belonging to the clones $\mathsf{S}_1^n, \mathsf{S}_{12}^n, \mathsf{S}_{11}^n, \mathsf{S}_{10}^n$ follow from the preceding results by duality. □

Taking inspiration from bases of relational clones given in [8, Table 1, p. 61], it would be desirable to prove that at least $\mathsf{L}, \mathsf{L}_0, \mathsf{L}_1, \mathsf{L}_2$ have weak base relations of the form $\mathrm{R}_{\mathsf{L}} = \Gamma_{\mathsf{L}}(\chi_2)$, $\{0\} \times \mathrm{R}_{\mathsf{L}}$, $\mathrm{R}_{\mathsf{L}} \times \{1\}$, $\{0\} \times \mathrm{R}_{\mathsf{L}} \times \{1\}$ (note that R_{L} consists of those eight quadruples in $\{0,1\}^4$ that have an even number of entries equal to 1). However, the following lemma shows that this hope is in vain.

Lemma 16. *$\{\{0\} \times \mathrm{R}_{\mathsf{L}}\}$, $\{\mathrm{R}_{\mathsf{L}} \times \{1\}\}$, and $\{\{0\} \times \mathrm{R}_{\mathsf{L}} \times \{1\}\}$ fail to be weak bases for L_0, L_1, and L_2, respectively. Instead, one may use $\rho_{\mathsf{L}_0} = \Gamma_{\mathsf{L}_0}(\chi_2)$, $\rho_{\mathsf{L}_1} = \Gamma_{\mathsf{L}_1}(\chi_2)$, $\rho_{\mathsf{L}_2} = \Gamma_{\mathsf{L}_2}(\chi_3)$ and $\rho_{\mathsf{L}_3} = \Gamma_{\mathsf{L}_3}(\chi_3)$ as irredundant weak base relations.*

Proof. According to [24, Table 1], the clones L_0 and L_1 have core size 2, L_2 and L_3 have core size 3. Hence Corollary 6 tells us that ρ_{L_a} where $a \in \{0,1,2,3\}$ are irredundant weak base relations of the respective clones; in particular these relations are generators for the relational clone $\operatorname{Inv} \mathsf{L}_a$. We now discuss the weak base failures.

By duality between L_1 and L_0, it is sufficient to show that $\{0\} \times \mathrm{R}_{\mathsf{L}}$ does not give a weak base of L_0. Knowing that ρ_{L_0} generates $\operatorname{Inv} \mathsf{L}_0$, if $\{0\} \times \mathrm{R}_{\mathsf{L}}$ were to form a weak base of L_0, then $\{0\} \times \mathrm{R}_{\mathsf{L}} \in [\{\rho_{\mathsf{L}_0}\}]_{\wedge,=}$ by Lemma 4, equivalently, by Theorem 3, $\operatorname{pPol}\{\rho_{\mathsf{L}_0}\} \subseteq \operatorname{pPol}\{\{0\} \times \mathrm{R}_{\mathsf{L}}\}$. However, $f \colon D \subseteq 2^3 \longrightarrow 2$ defined on $D = \{(0,0,0), (1,0,0), (1,0,1), (1,1,0), (1,1,1)\}$ as $f(x,y,z) = x \wedge y \wedge z$ for $x, y, z \in 2$ fails to preserve $\{0\} \times \mathrm{R}_{\mathsf{L}}$, because the following matrix $X \in 2^{5\times 3}$ has its three columns in $\{0\} \times \mathrm{R}_{\mathsf{L}}$, its rows in D, but the application of f produces a tuple outside $\{0\} \times \mathrm{R}_{\mathsf{L}}$:

$$X := \begin{pmatrix} 0 & 0 & 0 \\ 1 & 0 & 0 \\ 1 & 0 & 1 \\ 1 & 1 & 0 \\ 1 & 1 & 1 \end{pmatrix} \begin{matrix} \in D \\ \in D \\ \in D \\ \in D \\ \in D \end{matrix} \rightsquigarrow \begin{pmatrix} f(0,0,0) \\ f(1,0,0) \\ f(1,0,1) \\ f(1,1,0) \\ f(1,1,1) \end{pmatrix} = \begin{pmatrix} 0 \\ 0 \\ 0 \\ 0 \\ 1 \end{pmatrix} \notin \{0\} \times \mathrm{R}_{\mathsf{L}}$$

Adding the bottom row $(1,1,1)$ to X, we immediately observe that f does not preserve $\{0\} \times \mathrm{R}_{\mathsf{L}} \times \{1\}$ either. One can now check that f preserves $\rho_{\mathsf{L}_0} = \{0\} \times \sigma$, where $\sigma = \{\, (x_1, x_2, x_3) \in 2^3 \mid x_1 \oplus x_2 \oplus x_3 = 0 \}$ consists of all triples with an even number of values 1. For this it is sufficient to see that f preserves the relation σ, as $\rho_{\mathsf{L}_0} \in [\{\{0\}, \sigma\}]_{\wedge,=}$ and f clearly preserves $\{0\}$. We thus take an arbitrary matrix

$X \in 2^{3\times 3}$ with rows $z_1, z_2, z_3 \in D$ and columns $r_1, r_2, r_3 \in \sigma$. We have to prove that $f(X) := f \circ (r_1, r_2, r_3) = (f(z_1), f(z_2), f(z_3))$ belongs to σ. This is trivial whenever $\mathbf{0} \in \{r_1, r_2, r_3\}$, because then $f(X) = \mathbf{0}$. Hence, we suppose now that none of the columns equals $\mathbf{0}$. If $\mathbf{0} \notin \{z_1, z_2, z_3\}$, then all rows of the matrix begin with 1, wherefore $r_1 = \mathbf{1} \notin \sigma$, a contradiction. Consequently, there is $i \in \{1, 2, 3\}$ such that $z_i = \mathbf{0}$. Then, in order to have an even number of ones in each column and to avoid any columns being $\mathbf{0}$, we must have $r_1 = r_2 = r_3$, whence $f \circ (r_1, r_2, r_3) = r_1 \in \sigma$ by the idempotence of f. This shows that always $f(X) \in \sigma$, and thus f preserves σ and hence ρ_{L_0}. We therefore know now that $f \in \mathrm{pPol}\{\rho_{\mathsf{L}_0}\} \setminus \mathrm{pPol}\{\{0\} \times \mathrm{R}_{\mathsf{L}}\}$, which precludes $\{\{0\} \times \mathrm{R}_{\mathsf{L}}\}$ from being a weak base of L_0. By duality we obtain the analogous statement for L_1 and $\mathrm{R}_{\mathsf{L}} \times \{1\}$.

Similarly, if $\{\{0\} \times \mathrm{R}_{\mathsf{L}} \times \{1\}\}$ were a weak base for L_2, then we would have to have $\{0\} \times \mathrm{R}_{\mathsf{L}} \times \{1\} \in [\{\rho_{\mathsf{L}_2}\}]_{\wedge,=}$, for L_2 has core size 3 (cf. [24, Table 1]). Hence, from Theorem 3, it would follow that $\mathrm{pPol}\{\rho_{\mathsf{L}_2}\} \subseteq \mathrm{pPol}\{\{0\} \times \mathrm{R}_{\mathsf{L}} \times \{1\}\}$, which is again impossible using the same $f \notin \mathrm{pPol}\{\{0\} \times \mathrm{R}_{\mathsf{L}} \times \{1\}\}$ as above. Here again, we need to check that f preserves $\rho_{\mathsf{L}_2} = \Gamma_{\mathsf{L}_2}(\chi_3) = \left\{e_1^{(3)} \circ \beta, e_2^{(3)} \circ \beta, e_3^{(3)} \circ \beta, g \circ \beta\right\}$, where $\beta\colon 8 \longrightarrow \{0, 1\}^3$ is the bijection from the construction of χ_3 and $g(x, y, z) := x \oplus y \oplus z$ for $x, y, z \in 2$. It is easy to see that if $h, \hbar \in \mathcal{J}_2^{(3)} \cup \{g\}$ are any two ternary operations from L_2 such that $\{x \in 2^3 \mid h(x) = 0\} \subseteq \{x \in 2^3 \mid \hbar(x) = 0\}$, then $h = \hbar$. Let us now consider an arbitrary matrix $X \in 2^{8\times 3}$ such that all its columns belong to $\Gamma_{\mathsf{L}_2}(\chi_3)$ and its rows belong to D. The columns are therefore given as $h_j \circ \beta$ for certain $h_1, h_2, h_3 \in \mathsf{L}_2^{(3)}$, and for all $z \in 2^3$ we have $(h_1(z), h_2(z), h_3(z)) \in D$. Since $\mathbf{0} \in D$ is the only triple in D that begins with a 0, for every $z \in 2^3$ where $h_1(z) = 0$, we must also have $h_2(z) = h_3(z) = 0$. This implies that the inclusion $\{x \in 2^3 \mid h_1(x) = 0\} \subseteq \{x \in 2^3 \mid h_j(x) = 0\}$ holds for all $j \in \{1, 2, 3\}$, and therefore $h_j = h_1$, i.e., $h_1 = h_2 = h_3$. Thus, all columns of X are identical, and hence $f(X)$ equals its first column $h_1 \circ \beta \in \Gamma_{\mathsf{L}_2}(\chi_3) = \rho_{\mathsf{L}_2}$ by the idempotence of f.

The two technical preservation properties $f \in \mathrm{pPol}\{\Gamma_{\mathsf{L}_0}(\chi_2)\} \setminus \mathrm{pPol}\{\{0\} \times \mathrm{R}_{\mathsf{L}}\}$ and $f \in \mathrm{pPol}\{\Gamma_{\mathsf{L}_2}(\chi_3)\} \setminus \mathrm{pPol}\{\{0\} \times \mathrm{R}_{\mathsf{L}} \times \{1\}\}$, which underlie this proof, have also been formally verified using the SMT solver Z3 [30, 31]. The corresponding data files are available from [2]. □

Theorem 17. *For $F \in \{\mathsf{R}, \mathsf{M}, \mathsf{E}, \mathsf{V}\}$ there are single irredundant relations ρ_F such that $\{\rho_F\}$, $\{\{0\} \times \rho_F\}$, $\{\rho_F \times \{1\}\}$, and $\{\{0\} \times \rho_F \times \{1\}\}$ are (irredundant) weak bases of F, F_0, F_1 and F_2, respectively. For $F = \mathsf{L}$ a simple construction of this form fails. For $F \in \{\mathsf{I}, \mathsf{N}\}$ there are irredundant weak base relations ρ_F such that weak bases of F_0, F_1, F_2 can be obtained from them by extending the tuples in ρ_F with uniquely determined values. For $F = \mathsf{S}_0^n$ irredundant weak base relations can* *be constructed from* ρ_F *as* $\{0\} \times \rho_F$, $(\{0\} \times \rho_F) \cup \{\mathbf{1}\}$, $\{0\} \times ((\{0\} \times \rho_F) \cup \{\mathbf{1}\})$, *for* F_2, F_1 *and* F_0, *respectively; for* $F = \mathsf{S}_1^n$ *dual constructions apply.*

Proof. This follows from Lemmata 9–16. □

5 Relationships between Boolean weak bases

In [27] it was studied for which pairs F, G of Boolean clones we have $\bigcup \mathcal{L}_{\restriction F} \subseteq \bigcup \mathcal{L}_{\restriction G}$, that is, whether a particular weak base W_G of G is expressible as $W_G \in [W_F]_{\wedge,=}$ by a conjunctive formula (possibly including equality) from a weak base W_F of F, or not. It turns out that in many cases this is impossible [27, Figure 1, Table IV]. Nevertheless, in some of these impossible cases, the weak bases may be closely related in a slightly different way, and these relationships can still be exploited to obtain reductions for certain computational problems (such as counting), in which the parameter used for measuring the complexity does not change too much, e.g., linearly (by a constant factor) or by an additive constant. In many of these relationships we exploit existentially quantified variables whose values are additionally uniquely determined by the values of (some of) the non-quantified variables as in [26], which leads to parsimonious reductions where the size of the instance grows only linearly, using the same argument as given before Lemma 4. Thus, in many cases shown below the existence of such relationships can be considered as a consequence of the theory developed in [26], but the precise nature of the expressibility for the individual examples does not follow from there. The details of how relations can be expressed are important insofar as the relationships are particularly useful if the quantified variables are determined by an empty tuple of the non-quantified variables, that is, have a constant value (this has been called *frozen existential quantification* in [32]), or if they are determined by a single one of the non-quantified variables. We comment on these aspects for the individual examples presented subsequently.

To shorten formulations, we stipulate for relation symbols $R, R'_1, \ldots, R'_m$, tuples of variables $\boldsymbol{x}$, $\boldsymbol{z}_1, \ldots, \boldsymbol{z}_m$, sequences of variables $\boldsymbol{y}$ and $u_1, \ldots, u_n$, and constants $a_1, \ldots, a_n \in \{0,1\}$ that the expression

$$`\boldsymbol{x} \in R \overset{\circ}{\iff} \exists! \boldsymbol{y} \in 2\colon \bigwedge_{i=1}^{m} \boldsymbol{z}_i \in R'_i \wedge \bigwedge_{j=1}^{n} u_j = a_j'$$

is true if and only if the following chain of implications, wherein $\exists!$ denotes unique existence of elements (values being functionally determined by the value of $\boldsymbol{x}$), holds:

$$\boldsymbol{x} \in R \implies \exists! \boldsymbol{y} \in \{0,1\}\colon \bigwedge_{i=1}^{m} \boldsymbol{z}_i \in R'_i$$

$$\begin{aligned}
&\Longrightarrow \exists! \boldsymbol{y} \in \{0,1\} \colon \bigwedge_{i=1}^{m} \boldsymbol{z}_i \in R_i' \wedge \bigwedge_{j=1}^{n} u_j = a_j \\
&\Longrightarrow \exists \boldsymbol{y} \in \{0,1\} \colon \bigwedge_{i=1}^{m} \boldsymbol{z}_i \in R_i' \\
&\Longrightarrow \boldsymbol{x} \in R.
\end{aligned}$$

Likewise, we use '$\boldsymbol{x} \in R \overset{\circ}{\Longleftrightarrow} \exists! \boldsymbol{y} \in \{0,1\} \colon \boldsymbol{z} \in R'$' to denote

$$\boldsymbol{x} \in R \Longrightarrow \exists! \boldsymbol{y} \in \{0,1\} \colon \boldsymbol{z} \in R' \Longrightarrow \exists \boldsymbol{y} \in \{0,1\} \colon \boldsymbol{z} \in R' \Longrightarrow \boldsymbol{x} \in R.$$

Lemma 18. *For all $x_1, \dots, x_5 \in 2 = \{0,1\}$ we have*

$$\begin{aligned}
(x_1, \dots, x_4) \in \mathrm{R}_{\mathsf{E}} &\overset{\circ}{\Longleftrightarrow} \exists! u \in 2 \colon (u, x_1, \dots, x_4) \in \rho_{\mathsf{E}_0} \wedge u = 0, \\
&\overset{\circ}{\Longleftrightarrow} \exists! u \in 2 \colon (x_1, \dots, x_4, u) \in \rho_{\mathsf{E}_1} \wedge u = 1, \\
(x_1, \dots, x_5) \in \rho_{\mathsf{E}_0} &\overset{\circ}{\Longleftrightarrow} \exists! u \in 2 \colon (x_1, \dots, x_5, u) \in \rho_{\mathsf{E}_2} \wedge u = 1, \\
(x_1, \dots, x_5) \in \rho_{\mathsf{E}_1} &\overset{\circ}{\Longleftrightarrow} \exists! u \in 2 \colon (u, x_1, \dots, x_5) \in \rho_{\mathsf{E}_2} \wedge u = 0.
\end{aligned}$$

Proof. This follows because the relations appearing on the right-hand side of the equivalence $\overset{\circ}{\Longleftrightarrow}$ are a direct product of the respective ones on the left-hand side with a singleton set, e.g., $\rho_{\mathsf{E}_0} = \{0\} \times \mathrm{R}_{\mathsf{E}}$, $\rho_{\mathsf{E}_2} = \{0\} \times \rho_{\mathsf{E}_1}$, etc., cf. Lemma 11. □

The dual of this result is as follows:

Lemma 19. *For all $x_1, \dots, x_5 \in 2 = \{0,1\}$ we have*

$$\begin{aligned}
(x_1, \dots, x_4) \in \rho_{\mathsf{V}} &\overset{\circ}{\Longleftrightarrow} \exists! u \in 2 \colon (x_1, \dots, x_4, u) \in \rho_{\mathsf{V}_0} \wedge u = 0, \\
&\overset{\circ}{\Longleftrightarrow} \exists! u \in 2 \colon (u, x_1, \dots, x_4) \in \rho_{\mathsf{V}_1} \wedge u = 1, \\
(x_1, \dots, x_5) \in \rho_{\mathsf{V}_0} &\overset{\circ}{\Longleftrightarrow} \exists! u \in 2 \colon (u, x_1, \dots, x_5) \in \rho_{\mathsf{V}_2} \wedge u = 1, \\
(x_1, \dots, x_5) \in \rho_{\mathsf{V}_1} &\overset{\circ}{\Longleftrightarrow} \exists! u \in 2 \colon (x_1, \dots, x_5, u) \in \rho_{\mathsf{V}_2} \wedge u = 0.
\end{aligned}$$

Proof. We argue as for Lemma 18, but employ Lemma 12 instead of Lemma 11. □

For the group of 'monotone' clones we have the following result:

Lemma 20. *For all $x_1, x_2, x_3 \in \{0,1\}$ we have*

$$\begin{aligned}
(x_1, x_2) \in \mathrm{R}_{\mathsf{M}} &\overset{\circ}{\Longleftrightarrow} \exists! u \in 2 \colon (u, x_1, x_2) \in \rho_{\mathsf{M}_0} \wedge u = 0, \\
&\overset{\circ}{\Longleftrightarrow} \exists! u \in 2 \colon (x_1, x_2, u) \in \mathrm{R}_{\mathsf{M}_1} \wedge u = 1, \\
(x_1, x_2, x_3) \in \rho_{\mathsf{M}_0} &\overset{\circ}{\Longleftrightarrow} \exists! u \in 2 \colon (x_1, x_2, x_3, u) \in \rho_{\mathsf{M}_2} \wedge u = 1, \\
(x_1, x_2, x_3) \in \mathrm{R}_{\mathsf{M}_1} &\overset{\circ}{\Longleftrightarrow} \exists! u \in 2 \colon (u, x_1, x_2, x_3) \in \rho_{\mathsf{M}_2} \wedge u = 0.
\end{aligned}$$

Proof. The proof is analogous to Lemma 18, now using the representations as a direct product with a singleton set from Lemma 10. □

The same happens for the clones of subset preserving operations.

Lemma 21. *For all $x, y \in \{0,1\}$ we have*

$$\begin{aligned} x \in \rho_{\mathsf{R}} &\overset{\circ}{\Longleftrightarrow} \exists! u \in 2\colon (u,x) \in \rho_{\mathsf{R}_0} \wedge u = 0, \\ &\overset{\circ}{\Longleftrightarrow} \exists! u \in 2\colon (x,u) \in \rho_{\mathsf{R}_1} \wedge u = 1, \\ (x,y) \in \rho_{\mathsf{R}_0} &\overset{\circ}{\Longleftrightarrow} \exists! u \in 2\colon (x,y,u) \in \rho_{\mathsf{R}_2} \wedge u = 1, \\ (x,y) \in \rho_{\mathsf{R}_1} &\overset{\circ}{\Longleftrightarrow} \exists! u \in 2\colon (u,x,y) \in \rho_{\mathsf{R}_2} \wedge u = 0. \end{aligned}$$

Proof. The proof works as for Lemma 18, using the definitions from Lemma 9. □

We observe the very similar shape of the expressions used in Lemmata 18 to 21, which is due to the uniform construction of weak base relations guaranteed by Theorem 17. Thus, if these relationships are employed in a gadget reduction, the reductions will always be parsimonious, the number of variables will increase by 1 (constant value), and the size of the instance will grow linearly as the additional variable appears in every atom of the formula.

Lemma 22. *For $n \in \mathbb{N}$, $n \geq 2$, and $x_1, \ldots, x_{n+2} \in \{0,1\}$ we have*

$$\begin{aligned} (x_1,\ldots,x_{n+1}) \in \mathrm{R}_{\mathsf{S}_0^n} &\overset{\circ}{\Longleftrightarrow} \exists! u \in 2\colon (u,x_1,\ldots,x_{n+1}) \in \rho_{\mathsf{S}_{02}^n} \wedge u = 0, && (1)\\ &\Longleftrightarrow \exists u \in 2\colon ((u,x_1,\ldots,x_{n+1}) \in \mathrm{R}_{\mathsf{S}_{01}^n} \wedge u = 0) \mathbin{\dot\vee} && (2)\\ &\qquad ((u,x_1,\ldots,x_{n+1}) \in \mathrm{R}_{\mathsf{S}_{01}^n} \wedge \\ &\qquad (u,x_1,\ldots,x_{n+1}) = \mathbf{1}), \\ (x_1,\ldots,x_{n+2}) \in \rho_{\mathsf{S}_{02}^n} &\overset{[27]}{\Longleftrightarrow} (x_1,x_1,x_2,\ldots,x_{n+2}) \in \rho_{\mathsf{S}_{00}^n}, && (3)\\ &\Longleftrightarrow \exists u \in 2\colon ((x_1,u,x_2,\ldots,x_{n+2}) \in \rho_{\mathsf{S}_{00}^n} \wedge u = 0) \mathbin{\dot\vee} && (4)\\ &\qquad ((x_1,u,x_2,\ldots,x_{n+2}) \in \rho_{\mathsf{S}_{00}^n} \wedge \\ &\qquad (u,x_2,\ldots,x_{n+2}) = \mathbf{1} \wedge x_1 = 0), \\ (x_1,\ldots,x_{n+2}) \in \mathrm{R}_{\mathsf{S}_{01}^n} &\overset{\circ}{\Longleftrightarrow} \exists! u \in 2\colon (u,x_1,\ldots,x_{n+2}) \in \rho_{\mathsf{S}_{00}^n} \wedge u = 0. && (5) \end{aligned}$$

Proof. The first equivalence (1) holds because $\rho_{\mathsf{S}_{02}^n} = \{0\} \times \mathrm{R}_{\mathsf{S}_0^n}$, see Lemma 15. The second equivalence (2) holds since, by Lemma 15, $\mathrm{R}_{\mathsf{S}_{01}^n} = \rho_{\mathsf{S}_{02}^n} \cup \{\mathbf{1}\}$, and this union is disjoint as $\rho_{\mathsf{S}_{02}^n} = \{0\} \times \mathrm{R}_{\mathsf{S}_0^n}$, i.e., every tuple in $\rho_{\mathsf{S}_{02}^n}$ begins with 0. The projection of the tuples in $\rho_{S_{02}^n}$ completely describes $R_{S_0^n}$ as shown before, and the projection of $\mathbf{1}$ to its last $n+1$ places is again $\mathbf{1} \in (2^n \setminus \{\mathbf{0}\}) \times \{1\} = R_{S_0^n}$, cf. Lemma 15.

From Lemma 15 we also have $\rho_{S_{00}^n} = \{0\} \times R_{S_{01}^n}$, which directly explains the final equivalence (5). Therefore, if $(x_1, x_1, \ldots, x_{n+2}) \in \rho_{S_{00}^n}$, then $(x_1, \ldots, x_{n+2}) \in R_{S_{01}^n}$ and $x_1 = 0$. Hence, $(x_1, \ldots, x_{n+2}) \in R_{S_{01}^n} = \rho_{S_{02}^n} \cup \{\mathbf{1}\}$ and $(x_1, \ldots, x_{n+2}) \neq \mathbf{1}$ as $x_1 = 0$; consequently, $(x_1, \ldots, x_{n+2}) \in \rho_{S_{02}^n}$. In the other direction, if we assume that $(x_1, \ldots, x_{n+2}) \in \rho_{S_{02}^n} = \{0\} \times R_{S_0^n}$, then $x_1 = 0$ and $(x_1, \ldots, x_{n+2}) \in \rho_{S_{02}^n} \subseteq R_{S_{01}^n}$. Therefore, $(x_1, x_1, \ldots, x_{n+2}) \in \{0\} \times R_{S_{01}^n} = \rho_{S_{00}^n}$, and we have shown (3). We remark that equivalence (3) has also been stated earlier in the proof of [27, Lemma 8].

Since the union $R_{S_{01}^n} = \rho_{S_{02}^n} \cup \{\mathbf{1}\} = \left(\{0\} \times R_{S_0^n}\right) \cup \{\mathbf{1}\}$ is disjoint, there are two kinds of tuples in $\rho_{S_{00}^n} = \{0\} \times R_{S_{01}^n}$: the tuple $(0, 1, \ldots, 1)$, and all the remaining tuples, which are of the form $(0, 0, x_2, \ldots, x_{n+2}) = (x_1, u, x_2, \ldots, x_{n+2})$ with the condition $x_1 = u = 0$. Their projections obtained by removing the second entry, i.e., $(0, 1, \ldots, 1)$ and $(x_1, \ldots, x_{n+2}) = (0, x_2, \ldots, x_{n+2}) = (u, x_2, \ldots, x_{n+2})$, belong to the relation $\rho_{S_{02}^n} = \{0\} \times R_{S_0^n}$ because $\mathbf{1} \in R_{S_0^n}$ and $(u, x_2, \ldots, x_{n+2}) \in R_{S_{01}^n} \setminus \{\mathbf{1}\} = \rho_{S_{02}^n}$ since $u = 0$. Conversely, if $(x_1, \ldots, x_{n+2}) \in \rho_{S_{02}^n} = \{0\} \times R_{S_0^n}$, then $x_1 = 0$. Letting $u := 0 = x_1$ and using equivalence (3) that was demonstrated above, we then have $(x_1, u, x_2, \ldots, x_{n+2}) = (x_1, x_1, x_2, \ldots, x_{n+2}) \in \rho_{S_{00}^n}$, finishing the proof of (4). □

The equivalences (1), (3) and (5) can be used to obtain parsimonious reductions between computational problems; in the case of (3) the number of variables does not change between the instances, in the case of (1) and (5) it grows by 1 as only a single variable with a constant value of 0 throughout the whole instance needs to be introduced. In all three cases the size of the instance grows linearly as the length of each atom in the formula extends by a constant factor. From the equivalences (2) and (4) one can obtain reductions, in which one introduces a single fresh variable u throughout the instance and replaces each $R_{S_0^n}$-atom by the corresponding u-extended $R_{S_{01}^n}$-atom or $\rho_{S_{00}^n}$-atom as mentioned in (2) or (4), respectively. Each atom will thereby grow by a constant factor, hence the size of the instance in such a reduction will grow only linearly. According to the equivalences (2) and (4), this leads to an 'almost parsimonious' reduction, that is, the number of solutions increases by 1. For brevity, we shall explain this in the case of (2), the analysis for (4) being analogous. There is the possibility that the value of u in a solution to the $R_{S_{01}^n}$-instance is 1. Since this variable occurs in every atom, the first component of the tuples for each atom is equal to 1, and, according to (2), this can only occur if the whole tuple is $\mathbf{1}$, i.e., u and all the other variables have the value 1 (which is indeed a solution to the instance we have reduced to). In all the other solutions the value of u is 0, and, according to (2), these solutions correspond bijectively to those of the original $R_{S_0^n}$-instance. Therefore, the number of solutions of the $R_{S_{01}^n}$-instance is equal to the number of solutions of the $\mathrm{R}_{\mathsf{S}_0^n}$-instance increased by one, and in a reduction of a counting problem one needs to subtract 1 to obtain the answer to the original problem.

The dual of the preceding result is as follows, and an analogous analysis of the parameter growth in reductions can be performed:

Lemma 23. *For $n \in \mathbb{N}$, $n \geq 2$, and $x_1, \ldots, x_{n+2} \in \{0,1\}$ we have*

$$\begin{aligned}
(x_1, \ldots, x_{n+1}) \in \mathrm{R}_{\mathsf{S}_1^n} &\overset{\circ}{\iff} \exists! u \in 2\colon (u, x_1, \ldots, x_{n+1}) \in \rho_{\mathsf{S}_{12}^n} \wedge u = 1, \\
&\iff \exists u \in 2\colon ((u, x_1, \ldots, x_{n+1}) \in \mathrm{R}_{\mathsf{S}_{11}^n} \wedge u = 1) \dot{\vee} \\
&\qquad\qquad ((u, x_1, \ldots, x_{n+1}) \in \mathrm{R}_{\mathsf{S}_{11}^n} \wedge \\
&\qquad\qquad (u, x_1, \ldots, x_{n+1}) = \mathbf{0}), \\
(x_1, \ldots, x_{n+2}) \in \rho_{\mathsf{S}_{12}^n} &\iff (x_1, x_1, x_2, \ldots, x_{n+2}) \in \rho_{\mathsf{S}_{10}^n}, \\
&\iff \exists u \in 2\colon ((x_1, u, x_2, \ldots, x_{n+2}) \in \rho_{\mathsf{S}_{10}^n} \wedge u = 1) \dot{\vee} \\
&\qquad\qquad ((x_1, u, x_2, \ldots, x_{n+2}) \in \rho_{\mathsf{S}_{10}^n} \wedge \\
&\qquad\qquad (u, x_2, \ldots, x_{n+2}) = \mathbf{0} \wedge x_1 = 1), \\
(x_1, \ldots, x_{n+2}) \in \mathrm{R}_{\mathsf{S}_{11}^n} &\overset{\circ}{\iff} \exists! u \in 2\colon (u, x_1, \ldots, x_{n+2}) \in \rho_{\mathsf{S}_{10}^n} \wedge u = 1.
\end{aligned}$$

There is also an immediate relationship between the weak base relations for the clones of zero-separating functions of different degrees, which, in principle, was already observed in the proof of [27, Lemma 8]. However, in two of the expressions given there, unfortunately a variable is missing and hence variable identifications occur in the wrong places; the precise relationships are as follows:

Lemma 24 (cf. [27, Lemma 8]). *For $n \in \mathbb{N}$, $n \geq 2$, and any $x_1, \ldots, x_{n+3} \in \{0,1\}$ we have*

$$\begin{aligned}
(x_1, \ldots, x_{n+1}) \in \mathrm{R}_{\mathsf{S}_0^n} &\iff (x_1, x_1, x_2, \ldots, x_{n+1}) \in \mathrm{R}_{\mathsf{S}_0^{n+1}}, \\
(x_1, \ldots, x_{n+2}) \in \rho_{\mathsf{S}_{02}^n} &\iff (x_1, x_2, x_2, x_3, \ldots, x_{n+2}) \in \rho_{\mathsf{S}_{02}^{n+1}}, \\
(x_1, \ldots, x_{n+2}) \in \mathrm{R}_{\mathsf{S}_{01}^n} &\iff (x_1, x_2, x_2, x_3, \ldots, x_{n+2}) \in \mathrm{R}_{\mathsf{S}_{01}^{n+1}}, \\
(x_1, \ldots, x_{n+3}) \in \rho_{\mathsf{S}_{00}^n} &\iff (x_1, x_2, x_3, x_3, x_4, \ldots, x_{n+3}) \in \rho_{\mathsf{S}_{00}^{n+1}}.
\end{aligned}$$

Proof. Since for $n \geq 2$ the relation $2^n \setminus \{\mathbf{0}\}$ lies at the heart of the construction of $\mathrm{R}_{\mathsf{S}_0^n} = (2^n \setminus \{\mathbf{0}\}) \times \{1\}$, and thus of the remaining weak base relations, the key fact to observe for the four equivalences is that for all $x_1, \ldots, x_n \in \{0,1\}$ we have $(x_1, x_2, \ldots, x_n) = \mathbf{0}$ iff $(x_1, x_1, x_2, \ldots, x_n) = \mathbf{0}$. Its contrapositive directly implies the first two equivalences because $R_{S_0^n} = (2^n \setminus \{\mathbf{0}\}) \times \{1\}$ and $\rho_{S_{02}^n} = \{0\} \times (2^n \setminus \{\mathbf{0}\}) \times \{1\}$, cf. Lemma 15. Adding the tuple $\mathbf{1}$ in the form $R_{S_{01}^n} = \rho_{S_{02}^n} \cup \{\mathbf{1}\}$ is compatible with the variable identification relating $\rho_{S_{02}^{n+1}}$ and $\rho_{S_{02}^n}$, which then shows the third equivalence. Now the fourth equivalence follows immediately from the third one since by Lemma 15 we have $\rho_{S_{00}^n} = \{0\} \times R_{S_{01}^n}$. □

By duality we also obtain the respective relationships between clones of one-separating operations.

Lemma 25. *For $n \in \mathbb{N}$, $n \geq 2$, and any $x_1, \dots, x_{n+3} \in \{0, 1\}$ we have*

$$\begin{aligned}
(x_1, \dots, x_{n+1}) \in R_{S_1^n} &\iff (x_1, x_1, x_2, \dots, x_{n+1}) \in R_{S_1^{n+1}}, \\
(x_1, \dots, x_{n+2}) \in \rho_{S_{12}^n} &\iff (x_1, x_2, x_2, x_3, \dots, x_{n+2}) \in \rho_{S_{12}^{n+1}}, \\
(x_1, \dots, x_{n+2}) \in R_{S_{11}^n} &\iff (x_1, x_2, x_2, x_3, \dots, x_{n+2}) \in R_{S_{11}^{n+1}}, \\
(x_1, \dots, x_{n+3}) \in \rho_{S_{10}^n} &\iff (x_1, x_2, x_3, x_3, x_4, \dots, x_{n+3}) \in \rho_{S_{10}^{n+1}}.
\end{aligned}$$

We see that Lemmata 24 and 25 provide parsimonious reductions, in which the number of variables in the instance does not change; clearly, the size of the instance grows only by a constant factor as each atom in the formula grows slightly in length.

For the weak base relations of the essentially nullary clones the expressibility relationships are mainly contained in Lemma 13. We here simply make the nature of the functional dependence of all variables that were existentially quantified in Lemma 13 explicit (and for the sake of presentation we omit the symbols $\overset{\circ}{\iff}$ and $\exists!$ which follow implicitly). As noted before such relationships, when employed in a gadget reduction, will lead to parsimonious reductions.

Lemma 26. *For all $x_1, \dots, x_4 \in \{0, 1\}$ we have*

$$\begin{aligned}
(x_1, \dots, x_4) \in R_{I} &\iff (0, x_2 \wedge \neg x_1, x_3 \wedge \neg x_1, x_4 \wedge \neg x_1, x_1, x_2, x_3, x_4) \in \rho_{I_0}, \\
&\iff (x_1, x_2, x_3, x_4, x_1 \vee \neg x_4, x_2 \vee \neg x_4, x_3 \vee \neg x_4, 1) \in \rho_{I_1},
\end{aligned}$$

$$\begin{aligned}
&(x_1, \dots, x_8) \in \rho_{I_0} \\
&\quad \iff (x_1, \neg x_8, x_2, \neg x_7, x_3, \neg x_6, x_4, \neg x_5, x_5, \neg x_4, x_6, \neg x_3, x_7, \neg x_2, x_8, 1) \in \rho_{I_2}, \\
&(x_1, \dots, x_8) \in \rho_{I_1} \\
&\quad \iff (0, x_1, \neg x_7, x_2, \neg x_6, x_3, \neg x_5, x_4, \neg x_4, x_5, \neg x_3, x_6, \neg x_2, x_7, \neg x_1, x_8) \in \rho_{I_2};
\end{aligned}$$

$$\begin{aligned}
(x_1, \dots, x_4) \in \Gamma_{I_0}(\chi_2) &\iff (x_1, \neg x_4, x_2, \neg x_3, x_3, \neg x_2, x_4, 1) \in \Gamma_{I_2}(\chi_3) = \chi_3, \\
(x_1, \dots, x_4) \in \Gamma_{I_1}(\chi_2) &\iff (0, x_1, \neg x_3, x_2, \neg x_2, x_3, \neg x_1, x_4) \in \Gamma_{I_2}(\chi_3) = \chi_3.
\end{aligned}$$

Proof. This is straightforwardly verified using Lemma 13 or Appendix A. □

If one uses the last four equivalences of Lemma 26 in a reduction for a CSP or CSP counting problem, each variable that would be (uniquely) existentially quantified is either frozen or uniquely determined by a single non-quantified variable. It therefore suffices to introduce a companion variable for each variable occurring in the instance, thus doubling the number of variables of the instance (constant factor 2). Thereby also the size of the instance roughly doubles as each atom roughly gets twice as long.

Gadget reductions obtained from the first two equivalences of Lemma 26 are not as benign in terms of variables. Since the additional variables depend on combinations of two non-quantified variables, the number of variables needed may grow quadratically during the reduction. Since the size of each atom still at most doubles in the reduction, the growth in size is again bounded by the constant factor 2.

A possibility to express the weak base relation R_{N} in terms of $\mathrm{R}_{\mathsf{N}_2}$ (or ρ_{N_2}) and unique existential quantification has been fully described in Lemma 14. We note that also in this case the values of the existentially quantified variables are each functionally determined by a single one of the non-quantified ones. Thus, as above, in a reduction the growth in the number of variables and in the size of the instance can be bounded linearly by a constant factor of 2. Clearly, such a reduction will be parsimonious.

Recalling the weak base relations $\mathrm{R}_{\mathsf{D}} = \Gamma_{\mathsf{D}}(\chi_1)$, $\rho_{\mathsf{D}_1} = \Gamma_{\mathsf{D}_1}(\chi_2) = \chi_2$, and $\rho_{\mathsf{D}_2} = \Gamma_{\mathsf{D}_2}(\chi_3)$ from Section 4, we observe a comparable situation. We shall see that $\rho_{\mathsf{D}_1} = \chi_2 \in [\{\rho_{\mathsf{D}_2}\}]_{\wedge}$, wherefore in a reduction the number of variables will not change, but the size will double (cf. Lemma 27); moreover $\mathrm{R}_{\mathsf{D}} = \Gamma_{\mathsf{D}}(\chi_1)$ can be expressed using $\rho_{\mathsf{D}_1} = \chi_2$ with the help of frozen existential quantification, wherefore in a reduction the number of variables will grow by a constant value of 2 and, similarly to Lemmata 18 to 21, the growth in size can be bounded by a constant factor. The details of how the relations can be expressed are contained in the following lemma.

Lemma 27. *For all* $x_1, \ldots, x_4 \in \{0,1\}$ *we have*

$$(x_1,x_2,x_3,x_4) \in \rho_{\mathsf{D}_1} \iff (x_1,x_2,x_3,x_4,x_1,x_2,x_3,x_4) \in \rho_{\mathsf{D}_2},$$
$$(x_1,x_2) \in \mathrm{R}_{\mathsf{D}} \overset{\circ}{\iff} \exists! u,v \in 2\colon (u,x_1,x_2,v) \in \rho_{\mathsf{D}_1} \wedge u = 0 \wedge v = 1.$$

Proof. The clone D_2 is minimal and it is generated by the ternary Boolean majority operation μ [15, Figure 2]. The ternary part $\mathsf{D}_2^{(3)} = \mathcal{J}_2^{(3)} \cup \{\mu\}$ consists of only four functions, whence $\rho_{\mathsf{D}_2} = \Gamma_{\mathsf{D}_2}(\chi_3)$ contains exactly the value tuples of these four ternary operations. The variable identification stated in the lemma eliminates the value tuples of the projection $e_1^{(3)}$ and of μ, wherefore only the two binary projections forming $\chi_2 = \rho_{\mathsf{D}_1}$ remain.

Moreover, if from the value tuples of the two binary projections in χ_2 one removes the two constant coordinates, then one obtains the relation $\{(0,1),(1,0)\}$, containing the value tuples of the only two unary selfdual operations $e_1^{(1)}$ and $\neg$. The resulting binary relation therefore coincides with $\Gamma_{\mathsf{D}}(\chi_1) = \mathrm{R}_{\mathsf{D}}$. □

The last remaining group of clones are the five affine linear clones. The possibility to express invariant relations, in particular weak base relations, in terms of others using formulæ with uniquely determined existentially quantified variables follows from [26, Theorem 15]. The following lemma lists a concrete expressibility result, allowing reductions from $\mathrm{R}_{\mathsf{L}} = \Gamma_{\mathsf{L}}(\chi_2)$ to each of $\rho_{\mathsf{L}_0} = \Gamma_{\mathsf{L}_0}(\chi_2)$, $\rho_{\mathsf{L}_1} = \Gamma_{\mathsf{L}_1}(\chi_2)$ and $\rho_{\mathsf{L}_3} = \Gamma_{\mathsf{L}_3}(\chi_3)$, and from there to $\rho_{\mathsf{L}_2} = \Gamma_{\mathsf{L}_2}(\chi_3)$. These relations form weak bases according to the minimum core sizes of the respective clones and they coincide with the relations given in [24, Table 1] up to a permutation of variables.

Lemma 28. *For all* $x_1, \ldots, x_8 \in \{0,1\}$ *we have*

$$\begin{aligned}
(x_1,\ldots,x_4) \in \mathrm{R}_{\mathsf{L}} &\overset{\circ}{\iff} \exists! u,v \in 2\colon (u,x_1,x_2,v) \in \rho_{\mathsf{L}_0} \wedge (u,x_3,x_4,v) \in \rho_{\mathsf{L}_0} \wedge u = 0,\\
&\iff (0,x_1,x_2,x_1 \oplus x_2) \in \rho_{\mathsf{L}_0} \wedge (0,x_3,x_4,x_1 \oplus x_2) \in \rho_{\mathsf{L}_0},\\
&\overset{\circ}{\iff} \exists! u,v \in 2\colon (v,x_1,x_2,u) \in \rho_{\mathsf{L}_1} \wedge (v,x_3,x_4,u) \in \rho_{\mathsf{L}_1} \wedge u = 1,\\
&\iff (1 \oplus x_1 \oplus x_2, x_1, x_2, 1) \in \rho_{\mathsf{L}_1} \wedge (1 \oplus x_1 \oplus x_2, x_3, x_4, 1) \in \rho_{\mathsf{L}_1},\\
&\overset{\circ}{\iff} \exists! x_1', x_2', x_3', x_4' \in 2\colon (x_1,x_2,x_3,x_4,x_4',x_3',x_2',x_1') \in \rho_{\mathsf{L}_3},\\
&\iff (x_1,x_2,x_3,x_4,\neg x_4,\neg x_3,\neg x_2,\neg x_1) \in \rho_{\mathsf{L}_3};
\end{aligned}$$

moreover

$$\begin{aligned}
(x_1,\ldots,x_4) \in \rho_{\mathsf{L}_0} &\overset{\circ}{\iff} \exists! u, x_2', x_3', x_4' \in 2\colon (x_1,x_2,x_3,x_4,x_4',x_3',x_2',u) \in \rho_{\mathsf{L}_2} \wedge u = 1,\\
&\iff (x_1,x_2,x_3,x_4,\neg x_4,\neg x_3,\neg x_2,1) \in \rho_{\mathsf{L}_2},\\
(x_1,\ldots,x_4) \in \rho_{\mathsf{L}_1} &\overset{\circ}{\iff} \exists! u, x_2', x_3', x_4' \in 2\colon (u,x_3',x_2',x_1',x_1,x_2,x_3,x_4) \in \rho_{\mathsf{L}_2} \wedge u = 0,\\
&\iff (0,\neg x_3,\neg x_2,\neg x_1,x_1,x_2,x_3,x_4) \in \rho_{\mathsf{L}_2},
\end{aligned}$$

and

$$\begin{aligned}
(x_1,\ldots,x_8) \in \rho_{\mathsf{L}_3} &\overset{\circ}{\iff} \exists! u,u',v,w \in 2\colon (w,x_1,x_2,u,u',x_7,x_8,v) \in \rho_{\mathsf{L}_2} \wedge w = 0 \wedge\\
&\qquad\qquad (w,x_3,x_4,u,u',x_5,x_6,v) \in \rho_{\mathsf{L}_2} \wedge v = 1,\\
&\iff (0,x_1,x_2,x_1 \oplus x_2, 1 \oplus x_1 \oplus x_2, x_7, x_8, 1) \in \rho_{\mathsf{L}_2} \wedge\\
&\qquad (0,x_3,x_4,x_1 \oplus x_2, 1 \oplus x_1 \oplus x_2, x_5, x_6, 1) \in \rho_{\mathsf{L}_2}.
\end{aligned}$$

Proof. The expressibility results can be computed by inspecting the matrix representations of the relations (the relationship between ρ_{L_3} and ρ_{L_2} has been verified previously in Example 1); the unique values of the existentially quantified variables have in each case been shown by giving a functional dependence below the quantified formula, the uniqueness itself is a consequence of the fact that after identifying variables in the relations on the right-hand side, the number of remaining tuples is the same as for the relation on the left-hand side.

To provide an idea how the given formulæ arise, we consider the following description of the weak base relations occurring in the lemma:

$$\begin{aligned}
\mathrm{R}_{\mathsf{L}} &= \left\{ (x_1,\dots,x_4) \in 2^4 \,\middle|\, x_1 \oplus \cdots \oplus x_4 = 0 \right\},\\
\rho_{\mathsf{L}_0} &= \left\{ (x_1,\dots,x_4) \in 2^4 \,\middle|\, x_1 = 0 \wedge x_2 \oplus x_3 \oplus x_4 = 0 \right\},\\
\rho_{\mathsf{L}_1} &= \left\{ (x_1,\dots,x_4) \in 2^4 \,\middle|\, x_1 \oplus x_2 \oplus x_3 = 1 \wedge x_4 = 1 \right\},\\
\rho_{\mathsf{L}_2} &= \left\{ (x_1,\dots,x_8) \in 2^8 \,\middle|\, x_1 = 0 \wedge x_2 \oplus x_3 \oplus x_4 = 0 \wedge x_8 = 1 \wedge \bigwedge_{i=2}^{4} x_{9-i} = \neg x_i \right\},\\
\rho_{\mathsf{L}_3} &= \left\{ (x_1,\dots,x_8) \in 2^8 \,\middle|\, x_1 \oplus \cdots \oplus x_4 = 0 \wedge \bigwedge_{i=1}^{4} x_{9-i} = \neg x_i \right\}.
\end{aligned}$$

Moreover, as every $a \in \{0,1\}$ satisfies $a \oplus a = 0$, clearly for all $x, y, u, v \in \{0,1\}$ the following equivalences hold

$$\begin{aligned}
x \oplus y \oplus u \oplus v = 0 &\iff x \oplus y = u \oplus v \iff \exists! z \in 2\colon x \oplus y = z \wedge u \oplus v = z,\\
x \oplus y \oplus u = 0 &\iff u = x \oplus y;
\end{aligned}$$

these can be used to rewrite the descriptions of the relations as stated in the lemma. □

From the expressions in Lemma 28 we obtain parsimonious reductions, where the size of the instance grows only linearly (bounded by a constant factor). The growth in the number of variables in each case is at most quadratic (as for the reductions arising from Lemma 26); for the reductions from R_{L} to ρ_{L_3}, and from ρ_{L_a} to ρ_{L_2} ($a \in \{0,1\}$), the growth in the number of variables is even only linear and can be bounded by a constant factor of 2 by introducing a companion with the negated value for each variable occurring in the instance.

We now turn to expressions relating different groups of clones.

Lemma 29 (cf. [27, Table III]). *For all $x_1, \dots, x_4 \in \{0,1\}$ we have*

$$\begin{aligned}
(x_1, x_2, x_3) \in \mathrm{R}_{\mathsf{M}_1} &\iff (x_1, x_2, x_3, x_3) \in \mathrm{R}_{\mathsf{S}_{01}^2},\\
(x_1, x_2, x_3, x_4) \in \rho_{\mathsf{M}_2} &\iff (x_1, x_2, x_3, x_4, x_4) \in \rho_{\mathsf{S}_{00}^2},
\end{aligned}$$

$$(x_1, x_2, x_3) \in \rho_{\mathsf{M}_0} \iff (x_3, x_2, x_1, x_1) \in \mathrm{R}_{\mathsf{S}^2_{11}},$$
$$(x_1, x_2, x_3, x_4) \in \rho_{\mathsf{M}_2} \iff (x_4, x_3, x_2, x_1, x_1) \in \rho_{\mathsf{S}^2_{10}}.$$

Proof. In principle these relationships are contained in Table III of [27], but the third and fourth equivalence are stated in that table with variable-permuted relations for S^2_{11} and S^2_{10} and with an incorrect substitution of variables into those. It is hence easier to verify these facts directly using the definitions of the relations given in Lemma 15 or in Appendix A. □

Lemma 30. *For all* $x_1, \ldots, x_5 \in \{0,1\}$ *we have*

$$\begin{aligned}(x_1, \ldots, x_5) \in \rho_{\mathsf{S}^2_{00}} &\overset{\circ}{\iff} \exists! x_2', x_3', x_4' \in 2\colon (x_1, x_4', x_3', x_2', x_2, x_3, x_4, x_5) \in \rho_{\mathsf{D}_2},\\ &\iff (x_1, \neg x_4, \neg x_3, \neg x_2, x_2, x_3, x_4, x_5) \in \rho_{\mathsf{D}_2},\\ (x_1, \ldots, x_5) \in \rho_{\mathsf{S}^2_{10}} &\overset{\circ}{\iff} \exists! x_2', x_3', x_4' \in 2\colon (x_5, x_3, x_4, x_2, x_2', x_4', x_3', x_1) \in \rho_{\mathsf{D}_2},\\ &\iff (x_5, x_3, x_4, x_2, \neg x_2, \neg x_4, \neg x_3, x_1) \in \rho_{\mathsf{D}_2}.\end{aligned}$$

Proof. It is immediate from the matrix representation of ρ_{D_2} that the projection of ρ_{D_2} to the coordinates specified in the lemma gives the relations $\rho_{\mathsf{S}^2_{00}}$ and $\rho_{\mathsf{S}^2_{10}}$, respectively. That the values of the existentially quantified variables are uniquely determined is a consequence of the fact that all relations in the lemma contain exactly four tuples; one observes during the projection that their values are determined as complements of other variables as indicated in the lemma. □

If one uses these expressions in a gadget reduction, it will be parsimonious. Moreover, similar to the last four equivalences of Lemma 26, the number of variables will grow only with a constant factor of at most 2; also the size of the instance will increase only by a factor of $8/5 < 2$. Hence number of variables and size will grow linearly.

Lemma 31. *For all* $x_1, x_2, x_3 \in \{0,1\}$ *we have*

$$\begin{aligned}(x_1, x_2) \in \mathrm{R}_{\mathsf{M}} &\iff (x_1, x_1, x_1, x_2) \in \mathrm{R}_{\mathsf{E}} &&\iff (x_1, x_1, x_2, x_2) \in \mathrm{R}_{\mathsf{E}},\\ &\iff (x_1, x_2, x_2, x_2) \in \rho_{\mathsf{V}} &&\overset{3}{\iff} (x_1, x_1, x_2, x_2) \in \rho_{\mathsf{V}},\\ (x_1, x_2, x_3) \in \rho_{\mathsf{M}_0} &\iff (x_2, x_3, x_3, x_3, x_1) \in \rho_{\mathsf{V}_0} &&\iff (x_2, x_2, x_3, x_3, x_1) \in \rho_{\mathsf{V}_0},\\ (x_1, x_2, x_3) \in \mathrm{R}_{\mathsf{M}_1} &\iff (x_1, x_1, x_1, x_2, x_3) \in \rho_{\mathsf{E}_1} &&\iff (x_1, x_1, x_2, x_2, x_3) \in \rho_{\mathsf{E}_1}.\end{aligned}$$

[3]In principle the relationship in the second line of Lemma 31 could be obtained from [27, Table III], but, contrary to the claim in [27], the formula presented there produces the dual order, i.e., indices 1 and 2 would have to be swapped in Table III. For the relationships between M_0 and V_0, and between M_1 and E_1, the article [27] uses different weak base relations with some incorrectly placed variable indices, hence our expressions cannot be derived from there.

Proof. All expressions are readily verified using the matrix representations of the relations involved; the first equivalence relating $\mathrm{R_M}$ and $\mathrm{R_E}$ has been shown in Example 1, the proof of the second one is similar. The equivalences in the second line thence follow by duality. All remaining equivalences are direct consequences of the ones in the first two lines, taking into account the definition of the involved relations and placing the variable with the constant value on both sides of the equivalence in the appropriate place. □

Again, Lemma 31 can be used to produce parsimonious reductions in which the number of variables stays identical and the size of the instance grows only linearly.

Lemma 32. *For all $x_1, \dots, x_6 \in \{0,1\}$ we have*

$$(x_1,\dots,x_4) \in \mathrm{R_E} \overset{\circ}{\iff} \exists! u, u', y_2, y_3 \in 2\colon (x_1, x_2, x_3, u, u', y_3, y_2, x_4) \in \Gamma_{\mathsf{I}}(\chi_3),$$

$$(x_1,\dots,x_5) \in \rho_{\mathrm{E}_0} \overset{\circ}{\iff} \exists! y_2, y_3, y_4, x_6, y_6, x_7, y_7, x_8, y_8, x_9, y_9 \in 2\colon$$
$$(x_1, x_2, x_6, x_7, x_8, x_3, y_4, x_9, y_9, x_4, y_3, y_8, y_7, y_6, y_2, x_5) \in \Gamma_{\mathsf{I}_0}(\chi_4),$$

$$(x_1,\dots,x_5) \in \rho_{\mathrm{E}_1} \overset{\circ}{\iff} \exists! y_2, y_3, y_4, x_6, y_6, x_7, y_7, x_8, y_8, x_9, y_9 \in 2\colon$$
$$(x_1, y_4, x_6, x_7, x_2, x_8, x_9, y_3, x_3, y_9, y_8, y_2, y_7, y_6, x_4, x_5) \in \Gamma_{\mathsf{I}_1}(\chi_4),$$

$$(x_1,\dots,x_6) \in \rho_{\mathrm{E}_2} \overset{\circ}{\iff} \exists! y_2, y_3, y_4, y_5, x_7, y_7, x_8, y_8, x_9, y_9, x_{10}, y_{10}, x_{11}, y_{11},$$
$$x_{12}, y_{12}, x_{13}, y_{13}, x_{14}, y_{14}, x_{15}, y_{15}, x_{16}, y_{16}, x_{17}, y_{17} \in 2\colon$$
$$(x_1, x_2, y_5, x_7, x_8, x_9, x_{10}, x_{11}, x_{12}, x_3, x_{13}, x_{14}, x_{15}, x_{16}, y_4, x_{17},$$
$$y_{17}, x_4, y_{16}, y_{15}, y_{14}, y_{13}, y_3, y_{12}, y_{11}, y_{10}, y_9, y_8, y_7, x_5, y_2, x_6)$$
$$\in \Gamma_{\mathsf{I}_2}(\chi_5).$$

The values of the existentially quantified variables are functionally determined by the values of x_2, x_3, x_4 in the case of $\mathrm{R_E}$ and ρ_{E_1}, and by x_3, x_4, x_5 in the other two cases.

Proof. These relationships follow immediately by deleting rows in the matrix representations of the relations $\Gamma_{\mathsf{I}}(\chi_3)$, $\Gamma_{\mathsf{I}_a}(\chi_4)$ $(a \in \{0,1\})$ and $\Gamma_{\mathsf{I}_2}(\chi_5)$ as explained in Example 1. That the values of the existentially quantified variables are uniquely determined by the variables on the left-hand side of the equivalence holds because all relations in the lemma contain precisely five tuples. It can be observed that the matrix rows corresponding to the variables x_2, x_3, x_4 and x_3, x_4, x_5, respectively, contain exactly five distinct triples, explaining the functional dependence of the quantified variables. □

From all expressions in Lemma 32 we can obtain parsimonious reductions in which the size of the instance grows linearly by a constant factor of at most 32/6.

The growth in the number of variables, however, is bounded by a cubic polynomial, not linearly. We forego explicitly listing the dual of Lemma 32, which would allow to construct reductions from dual Horn clones to clones generated by constants.

As the weak base relations for the monotone group of clones have a much simpler shape, one can observe that the degree of the polynomial bounding the growth of the number of variables in a direct reduction to the weak bases of the clones $\mathsf{I}, \mathsf{I}_0, \mathsf{I}_1, \mathsf{I}_2$ decreases in comparison to a two-step reduction via the Horn group of clones. This is a consequence of the following result.

Lemma 33. *For all* $x_1, \ldots, x_4 \in \{0,1\}$ *we have*

$$\begin{aligned}
(x_1, x_2) \in \mathrm{R}_{\mathsf{M}} &\overset{[27]}{\iff} (x_1, x_1, x_2, x_2) \in \mathrm{R}_{\mathsf{I}} = \Gamma_{\mathsf{I}}(\chi_2),\\
(x_1, x_2, x_3) \in \rho_{\mathsf{M}_0} &\overset{\circ}{\iff} \exists! u \in 2\colon (x_1, u, x_2, x_3) \in \Gamma_{\mathsf{I}_0}(\chi_2),\\
&\iff (x_1, x_2 \oplus x_3, x_2, x_3) \in \Gamma_{\mathsf{I}_0}(\chi_2),\\
(x_1, x_2, x_3) \in \mathrm{R}_{\mathsf{M}_1} &\overset{\circ}{\iff} \exists! u \in 2\colon (x_1, x_2, u, x_3) \in \Gamma_{\mathsf{I}_1}(\chi_2),\\
&\iff (x_1, x_2, \neg(x_1 \oplus x_2), x_3) \in \Gamma_{\mathsf{I}_1}(\chi_2),\\
(x_1, \ldots, x_4) \in \rho_{\mathsf{M}_2} &\overset{\circ}{\iff} \exists! y_2, y_3, x_5, y_5 \in 2\colon (x_1, x_2, x_5, x_3, y_3, y_5, y_2, x_4) \in \Gamma_{\mathsf{I}_2}(\chi_3),\\
&\iff (x_1, x_2, x_2 \oplus x_3, x_3, \neg x_3, \neg(x_2 \oplus x_3), \neg x_2, x_4) \in \Gamma_{\mathsf{I}_2}(\chi_3).
\end{aligned}$$

Proof. The first equivalence has already been observed in [27, Table III], the others are easily verified using the matrix representations of the relations, and the values of the uniquely determined variables have in each case been specified below the existentially quantified formula. One may observe that in all three cases at least one of the existentially quantified variables is functionally determined by a pair of non-quantified variables, hence if these expressions are used in a reduction the number of variables may grow quadratically, although the size clearly increases only linearly. □

Lemma 34. *There is a permutation* $\pi \in \mathrm{Sym}\{1, \ldots, 64\}$*—for example, one may take the cycle product* $\pi = (1{\cdot}57{\cdot}8) \circ (2{\cdot}22) \circ (3{\cdot}36) \circ (4{\cdot}50) \circ (5{\cdot}15) \circ (6{\cdot}29) \circ (7{\cdot}43)$*—such that for all elements* $x_1, \ldots, x_8 \in \{0,1\}$ *we have*

$$\begin{aligned}
(x_1, \ldots, x_8) \in \rho_{\mathsf{D}_2} &\overset{\circ}{\iff} \exists! x_9, y_9, x_{10}, y_{10}, x_{11}, y_{11}, x_{12}, y_{12} \in 2\colon\\
&\qquad (x_1, x_2, x_3, x_9, x_5, x_{10}, x_{11}, x_{12}, y_{12}, y_{11}, y_{10}, x_4, y_9, x_6, x_7, x_8) \in \Gamma_{\mathsf{I}_2}(\chi_4),\\
(x_1, \ldots, x_4) \in \mathrm{R}_{\mathsf{N}} &\overset{\circ}{\iff} \exists! x_5, y_5, x_6, y_6, x_7, y_7, x_8, y_8, x_9, y_9, x_{10}, y_{10} \in 2\colon\\
&\qquad (x_5, x_6, x_7, x_1, x_8, x_9, x_2, x_{10}, y_{10}, x_3, y_9, y_8, x_4, y_7, y_6, y_5) \in \Gamma_{\mathsf{I}}(\chi_4),\\
(x_1, \ldots, x_8) \in \rho_{\mathsf{N}_2} &\overset{\circ}{\iff} \exists! x_9, \ldots, x_{64} \in 2\colon (x_{\pi(1)}, \ldots, x_{\pi(64)}) \in \Gamma_{\mathsf{I}_2}(\chi_6).
\end{aligned}$$

In all three equivalences the values of the existentially quantified variables are determined as functions of x_2, x_3, x_4.

Proof. The result and its proof are very similar to Lemma 32. Note that the quatuorsexagenary relation $\Gamma_{\mathsf{I}_2}(\chi_6) = \chi_6$ consists exactly of the six value tuples of the senary projection functions, and its matrix representation lists each Boolean sextuple exactly once (that is, in a uniquely determined row). Thus, if for $1 \leq i \leq 8$ the i-th row of the matrix representation of ρ_{N_2} appears as the j-th row of χ_6 (for a unique value of $j \in \{1, \ldots, 64\}$), we set $\pi(j) := i$. This partial definition can be extended to a permutation π of $\{1, \ldots, 64\}$, filling up the not yet defined indices $j \in \{1, \ldots, 64\} \setminus \{\pi^{-1}(1), \ldots, \pi^{-1}(8)\}$ with the values $\pi(j) \in \{9, \ldots, 64\}$. One particular choice for π has been stated in the lemma. Since the relations ρ_{N_2} and χ_6 both contain exactly 6 tuples, the values of the variables $x_9, \ldots, x_{64}$ in the formula are always uniquely determined by the values of $x_1, \ldots, x_8$. In fact, the rows of the matrix belonging to x_2, x_3, x_4 contain six (respectively four) distinct triples; hence these variables suffice to functionally determine the values of the quantified ones. □

One may turn the expressions in Lemma 34 into parsimonious reductions where the size of the instance will grow only linearly, but the number of variables may have to be increased cubically (possibly including a large constant factor).

Lemma 35. *For all* $x_1, \ldots, x_8 \in \{0, 1\}$ *we have*

$$\begin{aligned}
(x_1, \ldots, x_4) \in \mathrm{R}_{\mathsf{L}} &\overset{\circ}{\iff} \exists! y_1, \ldots, y_4 \in 2\colon (x_2, y_4, y_3, x_1, y_1, x_3, x_4, y_2) \in \Gamma_{\mathsf{N}}(\chi_3),\\
(x_1, \ldots, x_8) \in \rho_{\mathsf{L}_3} &\overset{\circ}{\iff} \exists! x_9, y_9, x_{10}, y_{10}, x_{11}, y_{11}, x_{12}, y_{12} \in 2\colon\\
&\qquad (x_2, x_9, x_{10}, x_1, x_{11}, x_3, x_4, x_{12}, y_{12}, x_5, x_6, y_{11}, x_8, y_{10}, y_9, x_7) \in \Gamma_{\mathsf{N}_2}(\chi_4),\\
(x_1, \ldots, x_8) \in \rho_{\mathsf{L}_2} &\overset{\circ}{\iff} \exists! x_9, y_9, x_{10}, y_{10}, x_{11}, y_{11}, x_{12}, y_{12} \in 2\colon\\
&\qquad (x_1, x_9, x_{10}, x_2, x_{11}, x_3, x_4, x_{12}, y_{12}, x_5, x_6, y_{11}, x_7, y_{10}, y_9, x_8) \in \Gamma_{\mathsf{I}_2}(\chi_4),\\
&\iff (x_1, x_3 \nrightarrow x_4, x_4 \nrightarrow x_3, x_2, x_3 \wedge x_4, x_3, x_4,\\
&\qquad x_3 \vee x_4, \neg(x_3 \vee x_4), x_5, x_6, \neg(x_3 \wedge x_4),\\
&\qquad x_7, x_4 \to x_3, x_3 \to x_4, x_8) \in \Gamma_{\mathsf{I}_2}(\chi_4),\\
(x_1 \ldots, x_4) \in \rho_{\mathsf{L}_0} &\overset{\circ}{\iff} \exists! y_1, \ldots, y_4 \in 2\colon (x_1, y_4, y_3, x_2, y_2, x_3, x_4, y_1) \in \Gamma_{\mathsf{I}_0}(\chi_3),\\
&\iff (x_1, x_2 \wedge x_3, x_2 \nrightarrow x_3, x_2, x_3 \nrightarrow x_2, x_3, x_4, x_2 \vee x_3) \in \Gamma_{\mathsf{I}_0}(\chi_3),\\
(x_1, \ldots, x_4) \in \rho_{\mathsf{L}_1} &\overset{\circ}{\iff} \exists! y_1, \ldots, y_4 \in 2\colon (y_4, x_3, x_2, y_1, x_1, y_2, y_3, x_4) \in \Gamma_{\mathsf{I}_1}(\chi_3),\\
&\iff (x_2 \wedge x_3, x_3, x_2, x_2 \vee x_3, x_1, x_2 \to x_3, x_3 \to x_2, x_4) \in \Gamma_{\mathsf{I}_1}(\chi_3).
\end{aligned}$$

In the first equivalence, the values of the existentially quantified variables are determined as functions of x_1, x_2, x_3, *and in the second one as functions of* x_1, x_3, x_4.

Proof. The equivalences follow by deleting the indicated rows in the matrix representations of the relations appearing on the right-hand sides. The values of the existentially quantified variables are uniquely determined since the number of tuples in the relations on both sides of each equivalence coincide. The details of how exactly they are determined can be seen from the matrix representation. □

The expressions in Lemma 35 can be used to obtain parsimonious reductions, in which the size of the instances grows only linearly. However, if one reduces from R_{L} or from ρ_{L_3}, the number of variables may increase cubically, in the other three cases quadratically.

Lemma 36. *For all $x, y, u, v \in \{0,1\}$ we have*[4]

$$\begin{aligned}(u,x,y,v) \in \rho_{\mathsf{D}_1} &\iff (u,u,x,x,y,y,v,v) \in \rho_{\mathsf{L}_2},\\ (x,y) \in \mathrm{R}_{\mathsf{D}} &\iff (x,y,y,x,y,x,x,y) \in \rho_{\mathsf{L}_3},\\ (x) \in \rho_{\mathsf{R}} &\iff (x,x,x,x) \in \mathrm{R}_{\mathsf{L}},\\ (x,y) \in \rho_{\mathsf{R}_0} &\iff (x,y,y,x) \in \rho_{\mathsf{L}_0},\\ (x,y) \in \rho_{\mathsf{R}_1} &\iff (y,x,x,y) \in \rho_{\mathsf{L}_1}.\end{aligned}$$

Proof. This is immediately visible after identifying rows in the matrix representations of the relations, cf. Appendix A. □

Lemma 37. *For all $x, y \in \{0,1\}$ we have*[4]

$$\begin{aligned}(x) \in \rho_{\mathsf{R}} &\iff (x,x) \in \mathrm{R}_{\mathsf{M}},\\ (x,y) \in \rho_{\mathsf{R}_0} &\iff (x,y,y) \in \rho_{\mathsf{M}_0},\\ (x,y) \in \rho_{\mathsf{R}_1} &\iff (x,x,y) \in \mathrm{R}_{\mathsf{M}_1},\\ (x,y,z) \in \rho_{\mathsf{R}_2} &\iff (x,y,y,z) \in \rho_{\mathsf{M}_2}.\end{aligned}$$

Proof. Since the order relation $\mathrm{R}_{\mathsf{M}} = \leq_2$ is reflexive, identifying its variables produces the full set $\{0,1\} = \rho_{\mathsf{R}}$. The other equivalences are direct consequences of this, exploiting the uniform construction of the weak base relations. □

Lemma 38. *For all $x, y, z \in \{0,1\}$ we have*[4]

$$\begin{aligned}(x,y) \in \rho_{\mathsf{R}_1} &\iff (x,y,y) \in \mathrm{R}_{\mathsf{S}_0^2},\\ (x,y) \in \rho_{\mathsf{R}_0} &\iff (y,x,x) \in \mathrm{R}_{\mathsf{S}_1^2},\\ (x,y,z) \in \rho_{\mathsf{R}_2} &\iff (x,y,z,z) \in \rho_{\mathsf{S}_{02}^2} \iff (z,y,x,x) \in \rho_{\mathsf{S}_{12}^2}.\end{aligned}$$

Proof. The proof is again straightforward, cf. Appendix A. □

Lemma 39. *For all $x, y, z \in \{0,1\}$ we have*

$$\begin{aligned}(x) \in \rho_{\mathsf{R}} &\overset{\circ}{\iff} \exists! u \in 2\colon (x,u) \in \mathrm{R}_{\mathsf{D}},\\ &\iff (x, \neg x) \in \mathrm{R}_{\mathsf{D}},\\ (x,y,z) \in \rho_{\mathsf{R}_2} &\overset{\circ}{\iff} \exists! u \in 2\colon (x,y,u,z) \in \rho_{\mathsf{D}_1},\\ &\iff (x,y,\neg y,z) \in \rho_{\mathsf{D}_1},\\ &\overset{4}{\iff} (x,x,z,z) \in \rho_{\mathsf{D}_1} \iff (x,z) \in \mathrm{R}_{\mathsf{R}_2}.\end{aligned}$$

Proof. These variable projections are evident by taking a quick glance at the relations $\mathrm{R}_{\mathsf{D}} = \Gamma_{\mathsf{D}}(\chi_1)$, $\rho_{\mathsf{D}_1} = \Gamma_{\mathsf{D}_1}(\chi_2) = \chi_2$ and $\mathrm{R}_{\mathsf{R}_2}$, cf. Appendix A. □

The equivalences in Lemmata 36 to 39 can be exploited to obtain parsimonious reductions where the number of variables does not change (for Lemmata 36 up to 38) or at most doubles (for the reductions based on the existentially quantified expressions from Lemma 39). In any case this is a linear increase in variables. The size of the instance will scale by a constant factor, i.e., increase only linearly.

Lemma 40. *For all $n \in \mathbb{N}$, $n \geq 2$ there are indices $0 \leq i_0, i_1, \ldots, i_{n+1} < 2^n$ such that for all $x_0, \ldots, x_{n+1} \in \{0,1\}$ we have*

$$(x_0, \ldots, x_{n+1}) \in \mathrm{R}_{\mathsf{S}_{01}^n} \overset{\circ}{\iff} (\exists! y_i \in 2)_{i \in \{0,\ldots,2^n-1\} \setminus \{i_0,\ldots,i_{n+1}\}} \colon \boldsymbol{y} \in \Gamma_{\mathsf{V}_1}(\chi_n),$$

where $\boldsymbol{y}_{i_j} = x_j$ for $0 \leq j \leq n+1$ and $\boldsymbol{y}_i = y_i$ for $0 \leq i < 2^n$, $i \notin \{i_0, \ldots, i_{n+1}\}$. Here the values of the existentially quantified variables can always be expressed as a function of the values of $x_1, \ldots, x_n$.

Proof. The clone V_1 is generated by $\{\vee, c_1\}$, cf. [15, Figure 2]. Therefore, $\Gamma_{\mathsf{V}_1}(\chi_n)$ contains the value tuples of all n-ary term operations of the algebra $\langle\{0,1\}; \vee, c_1\rangle$, of which there are precisely 2^n: the constant n-ary function with value 1, and for each non-empty set $I \subseteq \{1, \ldots, n\}$ the function $(a_1, \ldots, a_n) \mapsto \bigvee_{i \in I} a_i$. Since $\mathrm{R}_{\mathsf{S}_{01}^n} = (\{0\} \times (2^n \setminus \{\mathbf{0}\}) \times \{1\}) \cup \{\mathbf{1}\}$ has also 2^n elements, the existentially quantified variables on the right-hand side are uniquely determined by the values of $x_0, \ldots, x_{n+1}$; in fact, as $x_{n+1} = 1$ for every tuple $(x_0, \ldots, x_{n+1}) \in \mathrm{R}_{\mathsf{S}_{01}^n}$, they are determined by the values of $x_0, \ldots, x_n$, and one can also see that even $x_1, \ldots, x_n$ suffice.

For every $j \in \{1, \ldots, n\}$ define $I_j = \{1, \ldots, n\} \setminus \{j\}$ and let $f_j \in \mathsf{V}_1^{(n)}$ be given by $f_j(a_1, \ldots, a_n) = \bigvee_{i \in I_j} a_i$ for all $a_1, \ldots, a_n \in \{0,1\}$. There is exactly one tuple

[4] In [27, Table III] very similar relationships are presented; however, different weak base relations, e.g., for $\mathsf{R}, \mathsf{R}_0, \mathsf{R}_1, \mathsf{R}_2$, are used there, so no direct comparison can be made.

$(a_1, \ldots, a_n) \in 2^n \setminus \{\mathbf{0}\}$ such that $f_j(a_1, \ldots, a_n) = 0$, namely $\boldsymbol{e}_j = (0, \ldots, 0, 1, 0, \ldots, 0)$ where the 1 occurs in the j-th position. Let $i_j \in \{0, \ldots, 2^n - 1\}$ be the coordinate of $\Gamma_{\mathsf{V}_1}(\chi_n)$ corresponding to this tuple, that is, $\beta(i_j) = \boldsymbol{e}_j$ or $i_j := \beta^{-1}(\boldsymbol{e}_j)$ in the terminology of Section 2. The projection of $\Gamma_{\mathsf{V}_1}(\chi_n)$ to $i_1, \ldots, i_n$ gives the relation $\{0,1\}^n \setminus \{\mathbf{0}\}$. Letting $i_0 := \beta^{-1}(\mathbf{0})$ and $i_{n+1} := \beta^{-1}(\mathbf{1})$ be the coordinates corresponding to $\mathbf{0}$ and $\mathbf{1}$, respectively, the projection of $\Gamma_{\mathsf{V}_1}(\chi_n)$ to $i_0, \ldots, i_{n+1}$ yields $\mathrm{R}_{\mathsf{S}_{01}^n}$. □

If Lemma 40 is used to produce a parsimonious reduction from $\mathrm{R}_{\mathsf{S}_{01}^n}$ to $\Gamma_{\mathsf{V}_1}(\chi_n)$ for some fixed $n \in \mathbb{N}$, $n \geq 2$, then the size s and the number k of variables of an instance will grow at most as follows: $s \mapsto \kappa_n \cdot s$ where $\kappa_n := \frac{2^n}{n+2}$, and $k \mapsto k + \lambda_n \cdot k^n$ where $\lambda_n := 2^n - (n+2)$. This is so because in each atom there are λ_n places with existentially quantified variables the values of which are determined by the values of the combination of those n variables that are placed in the positions $i_1, \ldots, i_n$ of the atom. For each of these k^n value combinations of the k variables (and each of the possibly λ_n distinct functional dependencies) we introduce a new variable in the reduction. It seems possible that the degree of the polynomial bound on the number of variables could be reduced further by investigating the specific λ_n functional dependencies; however, this appears to be simpler to do, if it is needed, in the case of a concrete value of n. We note also that the quantities λ_n and κ_n, although they grow exponentially fast with n, are constants for each individual reduction, as they depend only on the considered constraint language, not on the instance that is transformed.

Of course, a dual result can be formulated concerning a parsimonious reduction from $\mathrm{R}_{\mathsf{S}_{11}^n}$-instances to $\Gamma_{\mathsf{E}_0}(\chi_n)$-instances for $n \geq 2$. Note, however, that $\Gamma_{\mathsf{V}_1}(\chi_n)$ and $\Gamma_{\mathsf{E}_0}(\chi_n)$ only form weak bases if $n \geq 3$ since V_1 and E_0 have minimum core size 3 each. Thus, for reductions between weak bases, one would reduce from $\mathrm{R}_{\mathsf{S}_{01}^2}$ to $\mathrm{R}_{\mathsf{S}_{01}^3}$ and then to $\Gamma_{\mathsf{V}_1}(\chi_3)$, and likewise from $\mathrm{R}_{\mathsf{S}_{11}^2}$ to $\mathrm{R}_{\mathsf{S}_{11}^3}$ and then to $\Gamma_{\mathsf{E}_0}(\chi_3)$.

Lemma 41. *For $n \in \mathbb{N}$, $n \geq 2$ there are $0 \leq i_{-1}, i_0, i_1, \ldots, i_{n+1}, l_1, \ldots, l_{n+1} < 2^{n+1}$ such that for all $x_{-1}, x_0, \ldots, x_{n+1} \in \{0,1\}$ we have*

$$(x_{-1}, x_0, \ldots, x_{n+1}) \in \rho_{\mathsf{S}_{00}^n} \stackrel{\circ}{\Longleftrightarrow} (\exists! y_i \in 2)_{i \in \{0,\ldots,2^{n+1}-1\} \setminus \{i_{-1}, i_0, \ldots, i_{n+1}, l_1, \ldots, l_{n+1}\}} : \boldsymbol{y} \in \Gamma_{\mathsf{V}_2}(\chi_{n+1}),$$

where $\boldsymbol{y}_{i_j} = \boldsymbol{y}_{l_j} = x_j$ for $1 \leq j \leq n+1$, $\boldsymbol{y}_{i_{-1}} = x_{-1}$, $\boldsymbol{y}_{i_0} = x_0$ and $\boldsymbol{y}_i = y_i$ for $0 \leq i < 2^{n+1}$, $i \notin \{i_{-1}, i_0, \ldots, i_{n+1}, l_1, \ldots, l_{n+1}\}$. Here the values of the existentially quantified variables can always be expressed as a function of the values of $x_1, \ldots, x_n$.

Proof. The proof of this lemma is similar to Lemma 40, but a little more technical. The clone V_2 is generated by $\{\vee\}$, cf. [15, Figure 2]. Therefore, $\Gamma_{\mathsf{V}_2}(\chi_{n+1})$ contains the value tuples of all $(n+1)$-ary term operations of the algebra $\langle\{0,1\};\vee\rangle$, of which there are precisely $2^{n+1}-1$: namely for each non-empty subset $I \subseteq \{1,\ldots,n+1\}$ the function $(a_1,\ldots,a_{n+1}) \overset{f_I}{\mapsto} \bigvee_{i\in I} a_i$. Their value tuples $f_I \circ \beta$ form the columns in the relation $\Gamma_{\mathsf{V}_2}(\chi_{n+1})$ (cf. Section 2). It will be our goal to use a variable identification with respect to $\Gamma_{\mathsf{V}_2}(\chi_{n+1})$ that keeps the 2^n-1 value tuples of the functions f_I where $\emptyset \neq I \subseteq \{1,\ldots,n\}$ and the one where $I = \{1,\ldots,n+1\}$. Up to reordering of columns, we are going to find the $n+3$ rows of the relation $\rho_{\mathsf{S}_{00}^n}$ in positions $i_{-1}, i_0 \ldots, i_{n+1}$ among the rows of the identification minor of $\Gamma_{\mathsf{V}_2}(\chi_{n+1})$. Since the minor and $\rho_{\mathsf{S}_{00}^n}$ both contain 2^n tuples, the values of the variables in the other rows are uniquely determined by the variable values in positions $i_{-1}, i_0 \ldots, i_{n+1}$; moreover, since the values in positions i_{-1} and i_{n+1} are constant 0 and 1, respectively, the values of $x_0,\ldots,x_n$ are sufficient to determine the remaining (existentially quantified) variables. In fact, the variable values in position i_0 only serve to distinguish the two columns $f_{\{1,\ldots,n\}} \circ \beta$ and $f_{\{1,\ldots,n+1\}} \circ \beta$ of the minor, which have identical values in the existentially quantified places. Hence the variable values of $x_1,\ldots,x_n$ suffice to functionally determine the values of the existentially quantified variables y_i.

This explains the uniqueness of the existential quantifier. We are now giving the variable identification. Similar to Lemma 40, we set $i_j := \beta^{-1}(\boldsymbol{e}_j)$ and $l_j := \beta^{-1}(\boldsymbol{e}'_j)$ for $1 \leq j \leq n$, where $\boldsymbol{e}_j = (0,\ldots,0,1,0,\ldots,0,0)$ and $\boldsymbol{e}'_j = (0,\ldots,0,1,0,\ldots,0,1)$, and in both $(n+1)$-tuples the left-most 1 appears in position j. Additionally, we set $i_{-1} := \beta^{-1}(\boldsymbol{0})$, $i_0 := \beta^{-1}(\boldsymbol{e}_{n+1})$, $i_{n+1} := \beta^{-1}(\boldsymbol{1})$ and $l_{n+1} := \beta^{-1}((1,\ldots,1,0))$. For any $1 \leq j \leq n$ and each non-empty $I \subseteq \{1,\ldots,n+1\}$, we have $f_I(\boldsymbol{e}_j) = f_I(\boldsymbol{e}'_j)$ if and only if the implication $n+1 \in I \implies j \in I$ holds for I. These are the functions the value tuples of which survive the variable identification $\boldsymbol{y}_{i_j} = \boldsymbol{y}_{l_j} = x_j$ in $\Gamma_{\mathsf{V}_2}(\chi_{n+1})$. Since we are employing the variable identifications for all $1 \leq j \leq n$ simultaneously, the functions f_I (tuples $f_I \circ \beta$) that are not removed by the identification are those that satisfy all of these implications, i.e., those where $n+1 \in I \implies \{1,\ldots,n\} \subseteq I$ holds. These are exactly the 2^n-1 functions f_I where $\emptyset \neq I \subseteq \{1,\ldots,n\}$ and, in addition, $f_{\{1,\ldots,n+1\}}$. Since $n > 0$, none of these is $e_{n+1}^{(n+1)}$, hence they also satisfy $f_I(\boldsymbol{1}) = 1 = f_I((1,\ldots,1,0))$ and thus survive $\boldsymbol{y}_{i_{n+1}} = \boldsymbol{y}_{l_{n+1}} = x_{n+1}$, too.

Projecting these 2^n tuples $f_I \circ \beta$ to the indices $i_1,\ldots,i_{n+1}$ corresponds to restricting the f_I to $\{\boldsymbol{e}_1,\ldots,\boldsymbol{e}_n,\boldsymbol{1}\}$ and produces every tuple in $(2^n \setminus \{\boldsymbol{0}\}) \times \{1\} = \mathrm{R}_{\mathsf{S}_0^n}$, where $\boldsymbol{1}$ appears twice, from $f_{\{1,\ldots,n\}}$ and from $f_{\{1,\ldots,n+1\}}$. If we project instead to $i_0,\ldots,i_{n+1}$, i.e., if we restrict the f_I to $\{\boldsymbol{e}_1,\ldots,\boldsymbol{e}_{n+1},\boldsymbol{1}\}$, then we are able to separate the two tuples $\boldsymbol{1}$ and obtain $(\{0\} \times (2^n \setminus \{\boldsymbol{0}\}) \times \{1\}) \cup \{\boldsymbol{1}\} = \rho_{\mathsf{S}_{02}^n} \cup \{\boldsymbol{1}\} = \mathrm{R}_{\mathsf{S}_{01}^n}$. Finally, if we project to $i_{-1},\ldots,i_{n+1}$, we restrict to $\{\boldsymbol{0},\boldsymbol{e}_1,\ldots,\boldsymbol{e}_{n+1},\boldsymbol{1}\}$ and thus add another zero in the first coordinate. Hence we obtain $\{0\} \times \mathrm{R}_{\mathsf{S}_{01}^{n}} = \rho_{\mathsf{S}_{00}^{n}}$ as the projection of the given identification minor of $\Gamma_{\mathsf{V}_2}(\chi_{n+1})$ to the places $i_{-1}, \ldots, i_{n+1}$ as claimed. The uniqueness of the values of the existentially quantified variables follows since the relation $\rho_{\mathsf{S}_{00}^{n}}$ on the left-hand side of the equivalence and the minor on the right-hand side both contain exactly 2^n tuples. □

By duality one can formulate from Lemma 41 a corresponding expressibility result for $\rho_{\mathsf{S}_{10}^{n}}$ in terms of $\Gamma_{\mathsf{E}_2}(\chi_{n+1})$ for $n \geq 2$. Due to the uniquely existentially quantified variables in Lemma 41 and its dual, we hence obtain parsimonious reductions from $\rho_{\mathsf{S}_{00}^{n}}$ to $\Gamma_{\mathsf{V}_2}(\chi_{n+1})$ and from $\rho_{\mathsf{S}_{10}^{n}}$ to $\Gamma_{\mathsf{E}_2}(\chi_{n+1})$. For fixed $n \geq 2$, in such a reduction the size s and the number of variables k of an instance grow within the following bounds: $s \mapsto \hat{\kappa}_n \cdot s$ and $k \mapsto k + \hat{\lambda}_n \cdot k^n$ where $\hat{\kappa}_n := \frac{2^{n+1}}{n+3}$ and $\hat{\lambda}_n := 2^{n+1} - 2(n+2)$ (in fact, the value $\hat{\lambda}_n$ could be halved since the existentially quantified variables can be grouped in identical pairs). In any case, for the parsimonious reductions established by Lemmata 40 and 41, the growth in size is linearly bounded, the growth in the number k of variables may depend polynomially on k with degree at most $n \geq 2$.

We have summarised the reductions that were presented in this article in Figure 3, including information on the growth of the number of variables between instances. All reductions exhibit an (at most) linear growth in the size of the instances; all, except for one type, are parsimonious, and the exceptional type is almost so: the number of solutions increases by 1. We note that in many cases, the weak bases appearing in the individual reductions match, so that transitivity can be applied seamlessly. There are, however, a few connections where one needs to switch the used weak base relations, and this can be done with the help of Lemma 4. As a consequence of this lemma any two irredundant, proper, non-empty weak base relations of the same relational clone parsimoniously inter-reduce to each other with no change in the number of variables of the instances and a linear bound on the growth in size. An example where such a change is necessary would be that one can parsimoniously reduce from R_{L} to $\Gamma_{\mathsf{N}}(\chi_3)$ via the expression shown in Lemma 35, then one has to invoke Lemma 4 to reduce from $\Gamma_{\mathsf{N}}(\chi_3)$ to $\mathrm{R}_{\mathsf{N}} = \Gamma_{\mathsf{N}}(\chi_2)$ before one may continue to reduce from R_{N} to $\Gamma_{\mathsf{I}}(\chi_4)$ by the expression from Lemma 34.

6 Concluding remarks

When tasked with the problem to provide reductions between computational problems parametrised by relations that do not exhibit an a priori compatibility with the existential quantifier or the equality predicate, weak bases with irredundant non-empty proper relations have proven a useful method. This is explained by Lemma 4 since problems parametrised by a weak base of a relational clone reduce to any problem parametrised by any generating set of that relational clone, as long as the problem is compatible with conjunction. Therefore, for obtaining complexity classifications it is usually a good start to connect and compare problems parametrised by weak bases.

In the Boolean case singleton weak bases are available for each relational clone, see [24] and the results in the article at hand, e.g., Theorem 17 and Appendix A. A helpful initial source of information to establish reductions between the weak base relations is [27, Figure 1]. However, many more paths for reduction are possible if parsimonious reductions are an option for the problem under consideration—counting CSPs being the prime example, for those reductions can be produced from expressions with existentially quantified variables whose values are uniquely determined, see [26]. In this article we have examined in detail reductions that are possible along *all* covering edges of Post's lattice and a few additional direct relationships. These are illustrated in Figure 3. In almost all cases our reductions are parsimonious, as predicted by the results of [26] on unique primitive positive definability. The only exceptions are notably the reductions between problems parametrised by the weak base relations $\mathrm{R}_{\mathsf{S}_0^n}$ and $\mathrm{R}_{\mathsf{S}_{01}^n}$ for each $n \in \mathbb{N}$, $n \geq 2$, following from Lemma 22, and their duals, following from Lemma 23. Here the target clones S_{01}^n and S_{11}^n are indeed among the few clones marked as 'not $\exists!$-covered' in [26, Figure 1] where such a situation would be conceivable. In these exceptional cases our proposed reduction is 'almost parsimonious' in the sense that the number of solutions changes only by one. Other target clones where, according to [26, Figure 1], we might have run into similar troubles, concern the reductions from R_{M} to R_{E} and to ρ_{V}, those from R_{E} to ρ_{E_0}, and those from ρ_{V} to ρ_{V_1}. However, Lemmata 18, 19 and 31 show that for the weak bases studied in this article unique primitive positive definitions and hence parsimonious reductions exist, which confirms what can be expected from [27, Figure 1].

For each relationship between Boolean relational clones that was investigated, a careful analysis has been performed regarding the growth of parameters that have been used in the past as measurements of complexity, namely the size of the formula and the number of distinct variables it contains. The increase in size is always bounded by a constant factor, that is, linear, being a consequence of a unique primitive positive (or even purely conjunctive) definition used to construct the reduction, or following from the specific reductions derived from Lemmata 22 and 23. In more than three fifths of the presented reductions the number of variables increases by at most two, in about three quarters, it increases not more than linearly; in about one fourth of the studied cases, however, our reductions entail a quadratic, cubic (or possibly higher degree, cf. Lemmata 40 and 41) blow-up in the number of variables.

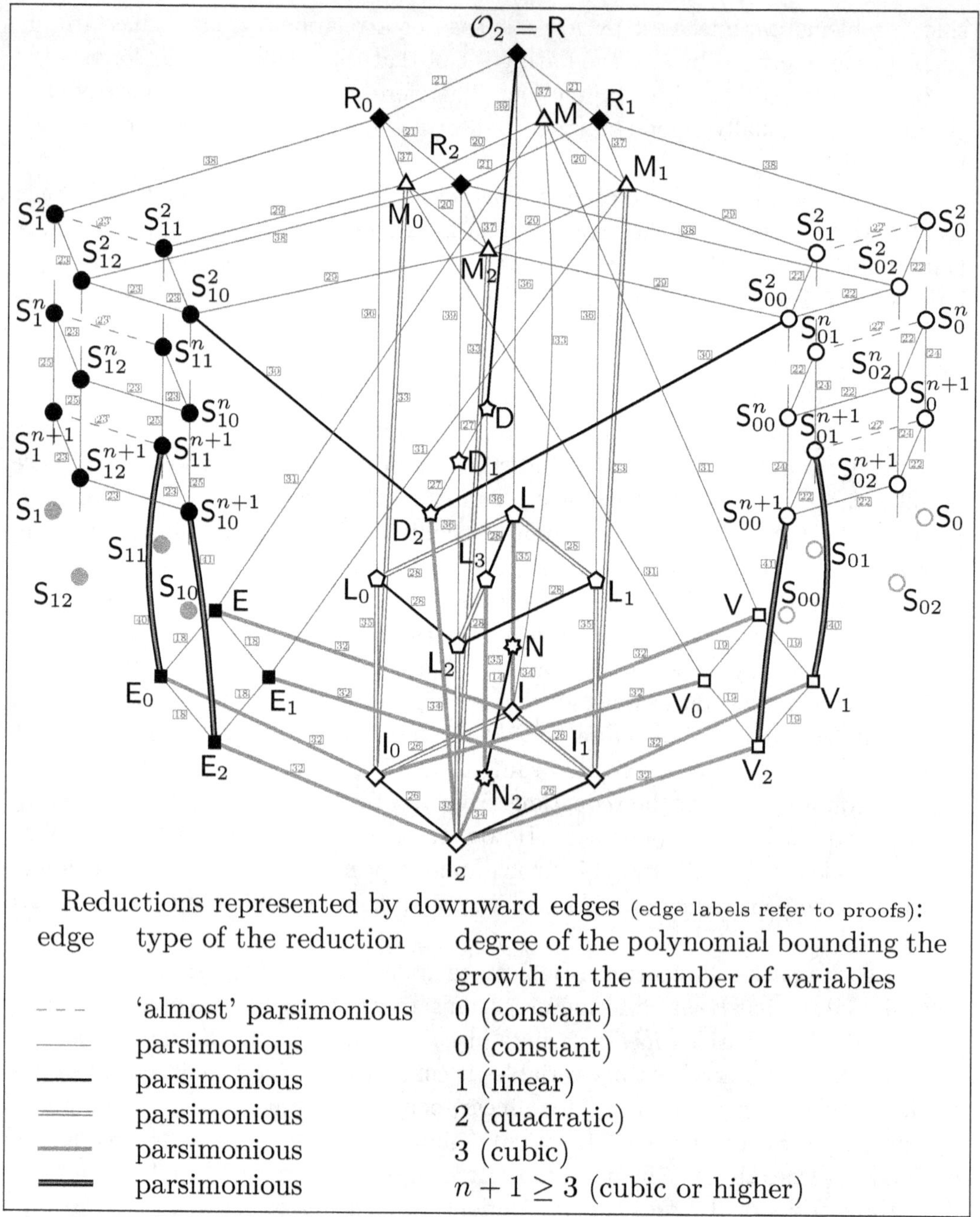

Figure 3: Reductions between weak bases of Boolean relational clones shown in this article: solid downward edges represent parsimonious reductions with a linear growth in size of the instance, while dashed edges are 'almost' parsimonious

We thus have identified these cases as the ones that in a complexity classification relying on reductions where the number of variables is not allowed to grow beyond linearly do not work 'out of the box' and hence necessitate special attention and possibly a separate argumentation.

References

[1] István Ágoston, János Demetrovics, and László Hannák. On the number of clones containing all constants (a problem of R. McKenzie). In *Lectures in universal algebra (Szeged, 1983)*, volume 43 of *Colloq. Math. Soc. János Bolyai*, pages 21–25. North-Holland, Amsterdam, 1986.

[2] Mike Behrisch. Verification of some Boolean partial polymorphisms [dataset]. *Zenodo*, 30 November 2021. 10.5281/zenodo.5745852.

[3] Mike Behrisch. Weak bases for Boolean relational clones revisited. In *2022 IEEE 52nd International Symposium on Multiple-Valued Logic—ISMVL 2022, Dallas, Texas, USA, 18–20 May 2022*, pages 68–73. IEEE Computer Soc., Los Alamitos, CA, May 2022. 10.1109/ISMVL52857.2022.00017.

[4] Mike Behrisch, Miki Hermann, Stefan Mengel, and Gernot Salzer. Give me another one! In Khaled M. Elbassioni and Kazuhisa Makino, editors, *Proceedings of the 26th International Symposium on Algorithms and Computation (ISAAC 2015), Nagoya (Japan)*, volume 9472 of *Lecture Notes in Comput. Sci.*, pages 664–676. Springer, Berlin, Heidelberg, November 2015. 10.1007/978-3-662-48971-0_56.

[5] Mike Behrisch, Miki Hermann, Stefan Mengel, and Gernot Salzer. As close as it gets. In Mohammad Kaykobad and Rossella Petreschi, editors, *Proceedings of the 10th International Workshop on Algorithms and Computation (WALCOM 2016), Lalitpur-Kathmandu (Nepal)*, volume 9627 of *Lecture Notes in Comput. Sci.*, pages 222–235. Springer, Berlin, Heidelberg, March 2016. 10.1007/978-3-319-30139-6_18.

[6] Mike Behrisch, Miki Hermann, Stefan Mengel, and Gernot Salzer. Minimal distance of propositional models. *Theory Comput. Syst.*, 63(6):1131–1184, 2019. 10.1007/s00224-018-9896-8.

[7] V. G. Bodnarčuk, Lev Arkaďevič Kalužnin, Victor N. Kotov, and Boris A. Romov. Теория Галуа для алгебр Поста. I, II [Galois theory for Post algebras. I, II]. *Kibernetika (Kiev)*, 5(3):1–10; ibid. 5(5):1–9, 1969.

[8] Elmar Böhler, Steffen Reith, Henning Schnoor, and Heribert Vollmer. Bases for Boolean co-clones. *Inform. Process. Lett.*, 96(2):59–66, October 2005. 10.1016/j.ipl.2005.06.003.

[9] Andrei A. Bulatov. The complexity of the counting constraint satisfaction problem. In Luca Aceto, Ivan Damgård, Leslie Ann Goldberg, Magnús M. Halldórsson, Anna Ingólfsdóttir, and Igor Walukiewicz, editors, *Automata, languages and programming, 35th International Colloquium, ICALP 2008, Reykjavik, Iceland, July 7–11 2008, Proceedings, Part I: Tack A: Algorithms, automata, complexity, and games*, volume 5125

of *Lecture Notes in Comput. Sci.*, pages 646–661. Springer, Berlin, 2008. 10.1007/978-3-540-70575-8_53.

[10] Andrei A. Bulatov. The complexity of the counting constraint satisfaction problem. *J. ACM*, 60(5):34:1–41, 2013. 10.1145/2528400.

[11] Andrei A. Bulatov. A dichotomy theorem for nonuniform CSPs. In *58th Annual IEEE Symposium on Foundations of Computer Science, FOCS 2017, Berkeley, CA, USA, 15–17 October 2017*, pages 319–330. IEEE Computer Soc., Los Alamitos, CA, October 2017. 10.1109/FOCS.2017.37.

[12] Andrei A. Bulatov and Víctor Dalmau. Towards a dichotomy theorem for the counting constraint satisfaction problem. *Inform. and Comput.*, 205(5):651–678, 2007. 10.1016/j.ic.2006.09.005.

[13] Miguel Couceiro, Lucien Haddad, and Victor Lagerkvist. Fine-grained complexity of constraint satisfaction problems through partial polymorphisms: a survey. In *2019 IEEE 49th International Symposium on Multiple-Valued Logic—ISMVL 2019, Fredericton, New Brunswick, Canada, 21–23 May 2019*, pages 170–175. IEEE Computer Soc., Los Alamitos, CA, May 2019. 10.1109/ISMVL.2019.00037.

[14] Miguel Couceiro, Lucien Haddad, and Karsten Schölzel. On the nonexistence of minimal strong partial clones. In *2017 IEEE 47th International Symposium on Multiple-Valued Logic—ISMVL 2017, Novi Sad, Serbia, 22–24 May 2017*, pages 82–87. IEEE Computer Soc., Los Alamitos, CA, May 2017. 10.1109/ISMVL.2017.43.

[15] Nadia Creignou and Heribert Vollmer. Boolean constraint satisfaction problems: When does Post's lattice help? In Nadia Creignou, Phokion G. Kolaitis, and Heribert Vollmer, editors, *Complexity of Constraints – An Overview of Current Research Themes [Result of a Dagstuhl Seminar]*, volume 5250 of *Lecture Notes in Comput. Sci.*, pages 3–37. Springer, Berlin, Heidelberg, 2008. 10.1007/978-3-540-92800-3_2.

[16] Martin Dyer and David Richerby. An effective dichotomy for the counting constraint satisfaction problem. *SIAM J. Comput.*, 42(3):1245–1274, 2013. 10.1137/100811258.

[17] David Scott Geiger. Closed systems of functions and predicates. *Pacific J. Math.*, 27(1):95–100, 1968. 10.2140/pjm.1968.27.95.

[18] Jurij Ivanovič Janov and Aľbert Abramovič Mučnik. О существовании k-значных замкнутых классов, не имеющих конечного базиса [On the existence of k-valued closed classes having no finite basis]. *Dokl. Akad. Nauk SSSR*, 127(1):44–46, 1959.

[19] Peter Jonsson, Victor Lagerkvist, Gustav Nordh, and Bruno Zanuttini. Strong partial clones and the time complexity of SAT problems. *J. Comput. System Sci.*, 84:52–78, March 2017. 10.1016/j.jcss.2016.07.008.

[20] Peter Jonsson, Victor Lagerkvist, and Biman Roy. Time complexity of constraint satisfaction via universal algebra. In Kim G. Larsen, Hans L. Bodlaender, and Jean-François Raskin, editors, *42nd International Symposium on Mathematical Foundations of Computer Science, MFCS 2017, Aalborg, Denmark, 21–25 August 2017*, volume 83 of *LIPIcs. Leibniz Int. Proc. Inform.*, pages 17:1–15. Schloß Dagstuhl. Leibniz-Zent. Inform., Wadern, 2017. 10.4230/LIPIcs.MFCS.2017.17.

[21] Peter Jonsson, Victor Lagerkvist, and Biman Roy. Fine-grained time complexity of

constraint satisfaction problems. *ACM Trans. Comput. Theory*, 13(1):2:1–32, March 2021. 10.1145/3434387.

[22] Peter Jonsson, Victor Lagerkvist, Johannes Schmidt, and Hannes Uppman. Relating the time complexity of optimization problems in light of the exponential-time hypothesis. In Erzsébet Csuhaj-Varjú, Martin Dietzfelbinger, and Zoltán Ésik, editors, *39th International Symposium on Mathematical Foundations of Computer Science, MFCS 2014, Budapest, Hungary, 25–29 August 2014. Proceedings, Part II*, volume 8635 of *Lecture Notes in Comput. Sci.*, pages 408–419. Springer, Heidelberg, 2014. 10.1007/978-3-662-44465-8_35.

[23] Peter Jonsson, Victor Lagerkvist, Johannes Schmidt, and Hannes Uppman. The exponential-time hypothesis and the relative complexity of optimization and logical reasoning problems. *Theoret. Comput. Sci.*, 892:1–24, November 2021. 10.1016/j.tcs.2021.09.006.

[24] Victor Lagerkvist. Weak bases of Boolean co-clones. *Inform. Process. Lett.*, 114(9):462–468, September 2014. 10.1016/j.ipl.2014.03.011.

[25] Victor Lagerkvist. Precise upper and lower bounds for the monotone constraint satisfaction problem. In Giuseppe F. Italiano, Giovanni Pighizzini, and Donald T. Sannella, editors, *40th International Symposium on Mathematical Foundations of Computer Science, MFCS 2015, Milan, Italy, 24–28 August 2015. Proceedings, Part I*, volume 9234 of *Lecture Notes in Comput. Sci.*, pages 357–368. Springer, Heidelberg, August 2015. 10.1007/978-3-662-48057-1_28.

[26] Victor Lagerkvist and Gustav Nordh. On the strength of uniqueness quantification in primitive positive formulas. In Peter Rossmanith, Pinar Heggernes, and Joost-Pieter Katoen, editors, *44th International Symposium on Mathematical Foundations of Computer Science, MFCS 2019, Aachen, Germany, 26–30 August 2019*, volume 138 of *LIPIcs. Leibniz Int. Proc. Inform.*, pages 36:1–15. Schloß Dagstuhl. Leibniz-Zent. Inform., Wadern, 2019. 10.4230/LIPIcs.MFCS.2019.36.

[27] Victor Lagerkvist and Biman Roy. The inclusion structure of Boolean weak bases. In *2019 IEEE 49th International Symposium on Multiple-Valued Logic—ISMVL 2019, Fredericton, New Brunswick, Canada, 21–23 May 2019*, pages 31–36. IEEE Computer Soc., Los Alamitos, CA, May 2019. 10.1109/ISMVL.2019.00014.

[28] Victor Lagerkvist and Biman Roy. Complexity of inverse constraint problems and a dichotomy for the inverse satisfiability problem. *J. Comput. System Sci.*, 117:23–39, May 2021. 10.1016/j.jcss.2020.10.004.

[29] Victor Lagerkvist and Magnus Wahlström. The power of primitive positive definitions with polynomially many variables. *J. Logic Comput.*, 27(5):1465–1488, July 2017. 10.1093/logcom/exw005.

[30] Microsoft Research. Z3 Theorem Prover, 2021. Available on-line from `https://github.com/z3prover/z3`, see also the description from `https://web.archive.org/web/20210119175613if_/https://rise4fun.com/Z3/tutorial/guide`.

[31] Leonardo de Moura and Nikolaj Bjørner. Z3: an efficient SMT solver. In Cartic R. Ramakrishnan and Jakob Rehof, editors, *International Conference on Tools and Al-*

gorithms for the Construction and Analysis of Systems (TACAS 2008), volume 4963 of *Lecture Notes in Comput. Sci.*, pages 337–340. Springer, Berlin, Heidelberg, March 2008. 10.1007/978-3-540-78800-3_24.

[32] Gustav Nordh and Bruno Zanuttini. Frozen Boolean partial co-clones. In *2009 IEEE 39th International Symposium on Multiple-Valued Logic—ISMVL 2009, Naha, Okinawa, Japan, 21–23 May 2009*, pages 120–125. IEEE Computer Soc., Los Alamitos, CA, 2009. 10.1109/ISMVL.2009.10.

[33] Emil Leon Post. *The Two-Valued Iterative Systems of Mathematical Logic*, volume 5 of *Annals of Mathematics Studies*. Princeton University Press, Princeton, N. J., 1941.

[34] Boris A. Romov. Алгебры частичных функций и их инварианты [The algebras of partial functions and their invariants]. *Kibernetika (Kiev)*, 17(2):1–11, March 1981.

[35] Boris A. Romov. The algebras of partial functions and their invariants. *Cybernetics*, 17(2):157–167, March 1981. 10.1007/BF01069627.

[36] Henning Schnoor and Ilka Schnoor. Enumerating all solutions for constraint satisfaction problems. In Wolfgang Thomas and Pascal Weil, editors, *Proceedings of the 24th Annual Symposium on Theoretical Aspects of Computer Science, STACS 2007, Aachen, Germany, 22–24 February 2007*, volume 4393 of *Lecture Notes in Comput. Sci.*, pages 694–705. Springer, Berlin, 2007. 10.1007/978-3-540-70918-3_59.

[37] Henning Schnoor and Ilka Schnoor. Partial polymorphisms and constraint satisfaction problems. In Nadia Creignou, Phokion G. Kolaitis, and Heribert Vollmer, editors, *Complexity of Constraints – An Overview of Current Research Themes [Result of a Dagstuhl Seminar]*, volume 5250 of *Lecture Notes in Comput. Sci.*, pages 229–254. Springer, Berlin, Heidelberg, 2008. 10.1007/978-3-540-92800-3_9.

[38] Ilka Schnoor. *The Weak Base Method for Constraint Satisfaction*. PhD dissertation, Gottfried Wilhelm Leibniz Universität Hannover, 2008.

[39] Dmitriy Zhuk. An algorithm for constraint satisfaction problem. In *2017 IEEE 47th International Symposium on Multiple-Valued Logic—ISMVL 2017, Novi Sad, Serbia, 22–24 May 2017*, pages 1–6. IEEE Computer Soc., Los Alamitos, CA, May 2017. 10.1109/ISMVL.2017.20.

[40] Dmitriy Zhuk. A proof of CSP dichotomy conjecture. In *58th Annual IEEE Symposium on Foundations of Computer Science, FOCS 2017, Berkeley, CA, USA, 15–17 October 2017*, pages 331–342. IEEE Computer Soc., Los Alamitos, CA, October 2017. 10.1109/FOCS.2017.38.

[41] Dmitriy Zhuk. A proof of the CSP dichotomy conjecture. *J. ACM*, 67(5):30:1–78, August 2020. 10.1145/3402029.

A Matrix representations of weak base relations

Here we show representations of Boolean weak base relations as matrices the *columns* of which are the tuples contained in the relation. Several relations of high arity used in connection with the clones $\mathsf{I}, \mathsf{I}_0, \mathsf{I}_1, \mathsf{I}_2$ generated by constants are not presented, but are certainly clear how to be constructed.

$$\mathrm{R}_{\mathsf{R}} = \left\{\begin{matrix} 01 \\ 01 \end{matrix}\right\} \qquad \mathrm{R}_{\mathsf{R}_0} = \{0\} \qquad \mathrm{R}_{\mathsf{R}_1} = \{1\} \qquad \mathrm{R}_{\mathsf{R}_2} = \left\{\begin{matrix} 0 \\ 1 \end{matrix}\right\}$$

$$\rho_{\mathsf{R}} = \{01\} \qquad \rho_{\mathsf{R}_0} = \left\{\begin{matrix} 00 \\ 01 \end{matrix}\right\} \qquad \rho_{\mathsf{R}_1} = \left\{\begin{matrix} 01 \\ 11 \end{matrix}\right\} \qquad \rho_{\mathsf{R}_2} = \left\{\begin{matrix} 00 \\ 01 \\ 11 \end{matrix}\right\}$$

$$\mathrm{R}_{\mathsf{M}} = \left\{\begin{matrix} 001 \\ 011 \end{matrix}\right\} \qquad \rho_{\mathsf{M}_0} = \left\{\begin{matrix} 000 \\ 001 \\ 011 \end{matrix}\right\} \qquad \mathrm{R}_{\mathsf{M}_1} = \left\{\begin{matrix} 001 \\ 011 \\ 111 \end{matrix}\right\} \qquad \rho_{\mathsf{M}_2} = \left\{\begin{matrix} 000 \\ 001 \\ 011 \\ 111 \end{matrix}\right\}$$

$$\mathrm{R}_{\mathsf{E}} = \left\{\begin{matrix} 00001 \\ 01001 \\ 10001 \\ 11101 \end{matrix}\right\} \qquad \rho_{\mathsf{E}_0} = \left\{\begin{matrix} 00000 \\ 00001 \\ 01001 \\ 10001 \\ 11101 \end{matrix}\right\} \qquad \rho_{\mathsf{E}_1} = \left\{\begin{matrix} 00001 \\ 01001 \\ 10001 \\ 11101 \\ 11111 \end{matrix}\right\} \qquad \rho_{\mathsf{E}_2} = \left\{\begin{matrix} 00000 \\ 00001 \\ 01001 \\ 10001 \\ 11101 \\ 11111 \end{matrix}\right\}$$

$$\rho_{\mathsf{V}} = \left\{\begin{matrix} 00001 \\ 01101 \\ 10101 \\ 11101 \end{matrix}\right\} \qquad \rho_{\mathsf{V}_0} = \left\{\begin{matrix} 00001 \\ 01101 \\ 10101 \\ 11101 \\ 00000 \end{matrix}\right\} \qquad \rho_{\mathsf{V}_1} = \left\{\begin{matrix} 11111 \\ 00001 \\ 01101 \\ 10101 \\ 11101 \end{matrix}\right\} \qquad \rho_{\mathsf{V}_2} = \left\{\begin{matrix} 11111 \\ 00001 \\ 01101 \\ 10101 \\ 11101 \\ 00000 \end{matrix}\right\}$$

$$\mathrm{R}_{\mathsf{L}} = \left\{\begin{matrix} 00001111 \\ 01101001 \\ 10100101 \\ 11000011 \end{matrix}\right\} \qquad \rho_{\mathsf{L}_0} = \left\{\begin{matrix} 0000 \\ 0110 \\ 1010 \\ 1100 \end{matrix}\right\} \qquad \rho_{\mathsf{L}_1} = \left\{\begin{matrix} 0011 \\ 0101 \\ 1001 \\ 1111 \end{matrix}\right\} \qquad \rho_{\mathsf{L}_2} = \left\{\begin{matrix} 0000 \\ 0011 \\ 0101 \\ 0110 \\ 1001 \\ 1010 \\ 1100 \\ 1111 \end{matrix}\right\}$$

$$\mathrm{R}_{\mathsf{D}} = \left\{\begin{matrix} 01 \\ 10 \end{matrix}\right\} \qquad \rho_{\mathsf{D}_1} = \left\{\begin{matrix} 00 \\ 01 \\ 10 \\ 11 \end{matrix}\right\} \qquad \rho_{\mathsf{D}_2} = \left\{\begin{matrix} 0000 \\ 0010 \\ 0100 \\ 0111 \\ 1000 \\ 1011 \\ 1101 \\ 1111 \end{matrix}\right\} \qquad \rho_{\mathsf{L}_3} = \left\{\begin{matrix} 00001111 \\ 00111100 \\ 01011010 \\ 01101001 \\ 10010110 \\ 10100101 \\ 11000011 \\ 11110000 \end{matrix}\right\}$$

$$\mathrm{R}_{\mathsf{N}} = \left\{\begin{matrix} 000111 \\ 010101 \\ 100011 \\ 110001 \end{matrix}\right\} \qquad \mathrm{R}_{\mathsf{N}_2} = \left\{\begin{matrix} 000111 \\ 010101 \\ 100011 \\ 110001 \\ 111000 \\ 101010 \\ 011100 \\ 001110 \end{matrix}\right\} \qquad \rho_{\mathsf{N}_2} = \left\{\begin{matrix} 000111 \\ 010101 \\ 100011 \\ 110001 \\ 001110 \\ 011100 \\ 101010 \\ 111000 \end{matrix}\right\}$$

$$\Gamma_{\mathsf{N}}(\chi_3) = \left\{\begin{matrix} 00001111 \\ 00101101 \\ 01001011 \\ 01101001 \\ 10000111 \\ 10100101 \\ 11000011 \\ 11100001 \end{matrix}\right\} \quad \Gamma_{\mathsf{N}_2}(\chi_4) = \left\{\begin{matrix} 00001111 \\ 00011110 \\ 00101101 \\ 00111100 \\ 01001011 \\ 01011010 \\ 01101001 \\ 01111000 \\ 10000111 \\ 10010110 \\ 10100101 \\ 10110100 \\ 11000011 \\ 11010010 \\ 11100001 \\ 11110000 \end{matrix}\right\}$$

$$\mathrm{R}_{\mathsf{I}} = \left\{\begin{matrix} 0001 \\ 0101 \\ 1001 \\ 1101 \end{matrix}\right\} \quad \mathrm{R}_{\mathsf{I}_0} = \left\{\begin{matrix} 001 \\ 010 \\ 011 \\ 000 \end{matrix}\right\} \quad \mathrm{R}_{\mathsf{I}_1} = \left\{\begin{matrix} 011 \\ 101 \\ 001 \\ 111 \end{matrix}\right\} \quad \mathrm{R}_{\mathsf{I}_2} = \left\{\begin{matrix} 100 \\ 010 \\ 001 \\ 011 \\ 101 \\ 110 \\ 000 \\ 111 \end{matrix}\right\}$$

$$\rho_{\mathsf{I}_0} = \left\{\begin{matrix} 0000 \\ 0010 \\ 0100 \\ 0110 \\ 1000 \\ 1010 \\ 1100 \\ 1110 \end{matrix}\right\} \quad \rho_{\mathsf{I}_1} = \left\{\begin{matrix} 0001 \\ 0011 \\ 0101 \\ 0111 \\ 1001 \\ 1011 \\ 1101 \\ 1111 \end{matrix}\right\} \quad \rho_{\mathsf{I}_2} = \left\{\begin{matrix} 0000 \\ 0001 \\ 0010 \\ 0011 \\ 0100 \\ 0101 \\ 0110 \\ 0111 \\ 1000 \\ 1001 \\ 1010 \\ 1011 \\ 1100 \\ 1101 \\ 1110 \\ 1111 \end{matrix}\right\}$$

$$\mathrm{R}_{\mathsf{S}_0^2} = \left\{\begin{matrix} 011 \\ 101 \\ 111 \end{matrix}\right\} \quad \rho_{\mathsf{S}_{02}^2} = \left\{\begin{matrix} 000 \\ 011 \\ 101 \\ 111 \end{matrix}\right\} \quad \mathrm{R}_{\mathsf{S}_{01}^2} = \left\{\begin{matrix} 0001 \\ 0111 \\ 1011 \\ 1111 \end{matrix}\right\} \quad \rho_{\mathsf{S}_{00}^2} = \left\{\begin{matrix} 0000 \\ 0001 \\ 0111 \\ 1011 \\ 1111 \end{matrix}\right\}$$

$$\mathrm{R}_{\mathsf{S}_1^2} = \left\{\begin{matrix} 001 \\ 010 \\ 000 \end{matrix}\right\} \quad \rho_{\mathsf{S}_{12}^2} = \left\{\begin{matrix} 111 \\ 001 \\ 010 \\ 000 \end{matrix}\right\} \quad \mathrm{R}_{\mathsf{S}_{11}^2} = \left\{\begin{matrix} 1110 \\ 0010 \\ 0100 \\ 0000 \end{matrix}\right\} \quad \rho_{\mathsf{S}_{10}^2} = \left\{\begin{matrix} 1111 \\ 1110 \\ 0010 \\ 0100 \\ 0000 \end{matrix}\right\}$$

$$\mathrm{R}_{\mathsf{S}_0^n} = \left\{\begin{matrix} \neq \mathbf{0} \\ 1\cdots1 \end{matrix}\right\} \quad \rho_{\mathsf{S}_{02}^n} = \left\{\begin{matrix} 0\cdots0 \\ \neq \mathbf{0} \\ 1\cdots1 \end{matrix}\right\} \quad \mathrm{R}_{\mathsf{S}_{01}^n} = \left\{\begin{matrix} 0\cdots01 \\ \neq \mathbf{0}\ 1 \\ 1\cdots11 \end{matrix}\right\} \quad \rho_{\mathsf{S}_{00}^n} = \left\{\begin{matrix} 0\cdots00 \\ 0\cdots01 \\ \neq \mathbf{0}\ 1 \\ 1\cdots11 \end{matrix}\right\}$$

$$\mathrm{R}_{\mathsf{S}_1^n} = \left\{\begin{matrix} \neq \mathbf{1} \\ 0\cdots0 \end{matrix}\right\} \quad \rho_{\mathsf{S}_{12}^n} = \left\{\begin{matrix} 1\cdots1 \\ \neq \mathbf{1} \\ 0\cdots0 \end{matrix}\right\} \quad \mathrm{R}_{\mathsf{S}_{11}^n} = \left\{\begin{matrix} 1\cdots10 \\ \neq \mathbf{1}\ 0 \\ 0\cdots00 \end{matrix}\right\} \quad \rho_{\mathsf{S}_{10}^n} = \left\{\begin{matrix} 1\cdots11 \\ 1\cdots10 \\ \neq \mathbf{1}\ 0 \\ 0\cdots00 \end{matrix}\right\}$$

Received

On Representation of Maximally Asymmetric Functions Based on Decision Diagrams

Shinobu Nagayama
Dept. of Computer and Network Eng., Hiroshima City University, Hiroshima, JAPAN

Tsutomu Sasao
Dept. of Computer Science, Meiji University, Kawasaki, JAPAN

Jon T. Butler
Dept. of Electr. and Comp. Eng., Naval Postgraduate School, Monterey, CA USA

Abstract

Maximally asymmetric functions are multiple-valued functions that have the maximum distance from their nearest symmetric functions, in terms of Hamming distance. Maximally asymmetric functions can be promising for applications such as radio communication and cryptography, due to randomness of the functions. To promote such application studies, benchmarks for maximally asymmetric functions in compact form are needed. However, few studies on benchmark generation or even on characteristics for maximally asymmetric functions have been reported. Thus, this paper investigates representation of maximally asymmetric functions based on decision diagrams. This paper begins with proposing a method to compute asymmetry of a given function easily, and then, derives a new characteristic of maximally asymmetric functions based on the computation method. Using the derived characteristic, we also propose a method to automatically generate benchmarks for maximally asymmetric functions represented by decision diagrams. By comparing sizes of different types of decision diagrams, we consider suitable representations for maximally asymmetric functions.

This paper is an extension of [24].

1 Introduction

It is hard to over-state the importance of symmetric functions in the study of logic functions. Symmetric functions appear in most, if not all, textbooks on logic design. Indeed, older textbooks even include a discussion of the synthesis of symmetric functions in the realization of the old-style contact switch network, (e.g., pp. 158-171 in [17]). The basic binary logic operations, two-variable AND and OR, realize symmetric functions. Symmetric functions occur in the sum (three-variable majority) and carry (three-variable exclusive OR) functions of the full adder. In an almost forgotten set of articles approximately 50 years ago, the articles [4, 12, 18, 34] exploited the surprising result that any binary function f can be realized by a single symmetric function in which certain variables are combined into a single variable of f. The significance of symmetric functions in cryptographic applications is investigated in [11, 19, 21, 27, 28]. In 1999, [25] introduced rotation symmetric functions, which represent a subset of symmetric functions. Such functions can be used in the rounds of efficient hashing algorithms.

A symmetric function is unchanged by any permutation of its variables. Therefore, in order to determine if a function is symmetric, one need examine only the function itself, and not any other functions. On the other hand, for maximally asymmetric functions, it is hard to determine if they are maximally asymmetric by just examining functions themselves. This is because an n-variable maximally asymmetric function is defined as a function as far away as possible from the set of n-variable symmetric functions. The descriptor "far", in this case, refers to the Hamming distance across the function values. For example, in the case of two-variable binary functions, the eight non-symmetric functions represent all of the maximally asymmetric functions. That is, all non-symmetric functions are at a distance 1 from some symmetric function. For larger n, there are many more maximally asymmetric than there are symmetric functions [30]. These observations suggest that maximally asymmetric functions are more complex than symmetric functions.

As for bent functions [26], with the aim to apply to encryption technology, their characteristic analyses and generation methods have been widely studied [16, 29, 32]. It is said that maximally asymmetric functions can be promising for applications such as CDMA communication and encryption [2, 5, 8, 13, 33], due to randomness of the functions [3, 9, 14, 30]. However, as far as we know, the usefulness of maximally asymmetric functions in such applications has not been shown, since their characteristics and generation methods have not been reported as much as ones for symmetric functions. Although the number of n-variable r-valued maximally asymmetric functions has been shown in [8, 9], their characteristics as well as generation methods are not sufficiently known yet. Thus, this paper shows a new characteristic of maximally asymmetric functions. It is based on a new method to compute asymmetry of a given function easily, which is proposed in this paper. By using the derived characteristic, we can determine if a given function is maximally asymmetric by just examining the function itself.

To promote studies on maximally asymmetric functions, providing their benchmarks in compact form is important. This paper focuses on the use of decision diagrams as a means to compactly represent maximally asymmetric functions. Although decision diagrams are well-known as a compact form for various functions, the size of decision diagrams varies significantly, depending on classes of functions as well as the order of variables. Thus, it is important to choose an appropriate decision diagram to a targeting class of functions. To do that, this paper shows sizes of different types of decision diagrams, MDD, EVMDD, BDD, and ZDD, for maximally asymmetric functions. This paper also proposes a method to randomly generate benchmarks for maximally asymmetric functions represented by decision diagrams. As far as we know, this is the first time to propose the method for benchmark generation of maximally asymmetric functions. The proposed method is based on the new characteristic of maximally asymmetric functions.

The rest of this paper is organized as follows: Section 2 shows some definitions for maximally asymmetric functions and decision diagrams. Section 3 shows a characteristic of maximally asymmetric functions derived from the results of [8, 9, 10]. Based on the derived characteristic, Section 4 presents a generation method of benchmarks for n-variable r-valued maximally asymmetric functions in decision diagram form. Section 5 shows sizes of decision diagrams for randomly generated maximally asymmetric functions, and Section 6 concludes the paper.

2 Preliminaries

This section shows definitions of maximally asymmetric functions [8] and decision diagrams.

2.1 Maximally Asymmetric Functions

Definition 1. *For an n-variable r-valued function $f(X_1, X_2, \ldots, X_n) : \{0, 1, \ldots, r-1\}^n \to \{0, 1, \ldots, r-1\}$, an assignment of values to the n variables is an **input vector**. When the r^n input vectors $(0, 0, \ldots, 0), (0, 0, \ldots, 1), \ldots, (r-1, r-1, \ldots, r-1)$ are applied to the input variables of the function f in ascending order, the vector of obtained function values is the **function vector** F.*

Definition 2. *The* ***Hamming distance*** $d(F, G)$ *between two vectors* F *and* G *of the same length is the number of positionwise different elements between* F *and* G.

Definition 3. *A* ***symmetric function*** f *satisfies*

$$f(X_1, X_2, \dots X_i, \dots, X_j, \dots, X_n) = f(X_1, X_2, \dots X_j, \dots, X_i, \dots, X_n)$$

for any pair of variables X_i *and* X_j. *In a symmetric function, function values are decided only by combinations of values assigned to* $X_1, X_2, \dots, X_n$. *This is also known as v-symmetry (variable-symmetry) [7]. Although there is another symmetry (vv-symmetry [8]), this paper focuses only on v-symmetry since techniques in this paper are expandable to vv-symmetry.*

Definition 4. *The* ***asymmetry*** *of a function* f *is the Hamming distance between* f *and its nearest symmetric function* g, *and is defined as*

$$\min_{g \in \Gamma} d(F, G),$$

where F *and* G *are function vectors of* f *and* g, *respectively, and* Γ *is the set of all symmetric functions.*

Definition 5. *An* n*-variable* r*-valued* ***maximally asymmetric function (MAF)*** *has the larg-est asymmetry among all* n*-variable* r*-valued functions.*

Example 1. *Table 1 shows truth tables of* f_1 *and* f_2, *three-variable three-valued asymmetric functions, and of* g_1 *and* g_2, *symmetric functions. In Table 1, input vectors are reordered and grouped into the same combinations of input values. As shown in Table 1, function values of* g_1 *and* g_2 *depend on* ***combinations of input values*** *rather than individual input vectors. On the other hand, function values of* f_1 *and* f_2 *are independent of combinations of input values. Since* g_1 *is the nearest symmetric function of* f_1 *in terms of the Hamming distance, the asymmetry of* f_1 *is 16. Although the Hamming distance between* g_1 *and* f_2 *is 27, it is not the asymmetry of* f_2 *since* g_1 *is not the nearest symmetric function of* f_2. *Its nearest function is* g_2, *and thus, the asymmetry of* f_2 *is 7. Since* f_1 *has the largest asymmetry among three-variable three-valued functions,* f_1 *is maximally asymmetric.* □

In this way, to compute the asymmetry of a given n-variable r-valued function and decide whether it is maximally asymmetric or not, we have to find its nearest symmetric function and confirm that its asymmetry is the largest among n-variable r-valued functions. If we follow only the definitions (i.e., if we do not use any characteristics of MAFs shown later), then it is not easy to decide whether a given function is maximally asymmetric or not.

X_1	X_2	X_3	f_1	g_1	f_2	g_2
0	0	0	1	1	0	0
1	1	1	2	2	1	1
2	2	2	0	0	2	2
0	0	1	0	1	0	2
0	1	0	1	1	2	2
1	0	0	2	1	2	2
0	0	2	1	0	2	1
0	2	0	2	0	1	1
2	0	0	0	0	1	1
0	1	1	0	1	0	0
1	0	1	1	1	2	0
1	1	0	2	1	0	0
0	2	2	0	1	0	0
2	0	2	1	1	0	0
2	2	0	2	1	0	0
1	1	2	0	0	1	1
1	2	1	1	0	1	1
2	1	1	2	0	1	1
1	2	2	2	0	1	1
2	1	2	0	0	1	1
2	2	1	1	0	2	1
0	1	2	1	2	1	0
0	2	1	2	2	0	0
1	0	2	1	2	1	0
1	2	0	2	2	0	0
2	0	1	0	2	0	0
2	1	0	0	2	1	0

Table 1: Asymmetric functions f_1 and f_2, and symmetric functions g_1 and g_2.

2.2 Decision Diagrams

Definition 6. *A **binary decision diagram (BDD)** [1, 6] is a rooted directed acyclic graph (DAG) representing a binary logic function. The BDD is obtained by recursively applying the Shannon expansion $f = \overline{x_i}f_0 + x_i f_1$ to the logic function, where $f_0 = f(x_i = 0)$, and $f_1 = f(x_i = 1)$. It consists of two terminal nodes representing function values 0 and 1 respectively, and nonterminal nodes representing input variables. Each nonterminal node has two unweighted outgoing edges, 0-edge and 1-edge, that correspond to the values of the input variables. Both terminal nodes have no outgoing edges. In this paper, a BDD means a reduced ordered BDD (ROBDD) that is obtained by fixing the order of variables in a BDD, and by applying the following two reduction rules:*

1. *Coalesce equivalent sub-graphs.*

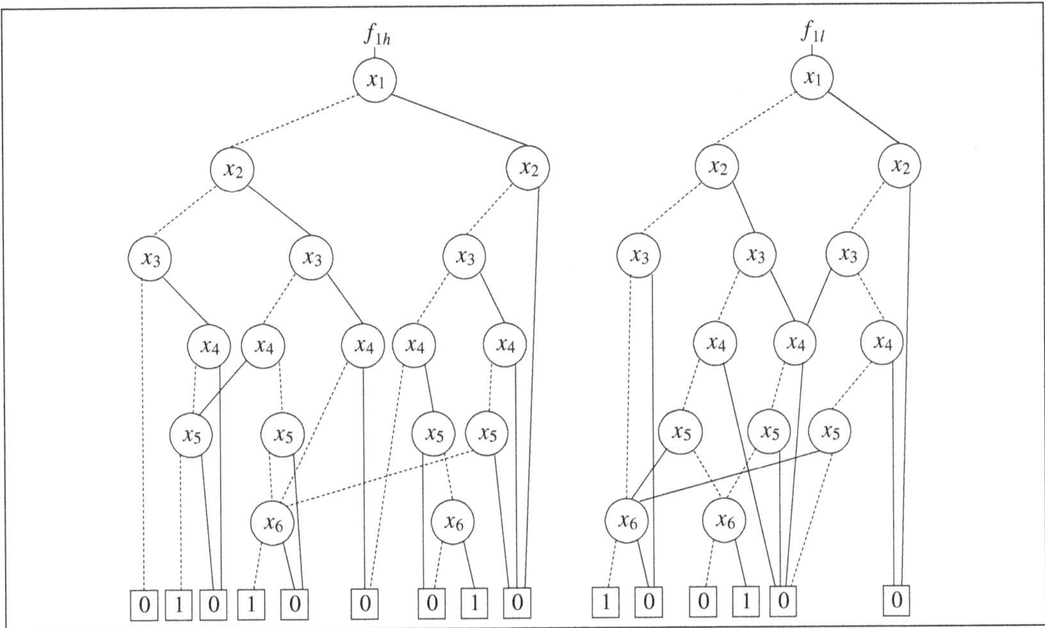

Figure 1: BDDs for $f_1 = (f_{1h}, f_{1l})$

2. *Delete nonterminal nodes v when both its outgoing edges point to the same node u , and redirect edges pointing to v to its child node u.*

Definition 7. *A* ***zero-suppressed binary decision diagram (ZDD)*** *[20] is a variant of a BDD. In ZDDs, the following reduction rules are used:*

1. *Coalesce equivalent sub-graphs.*

2. *Delete nonterminal nodes v whose 1-edge points to the terminal node representing 0, and redirect edges pointing to v to its child node u that is pointed by the 0-edge of v.*

BDDs and ZDDs represent multiple-valued functions by converting them into binary encoded multiple-output logic functions.

Example 2. *Figs. 1 and 2 show BDDs and ZDDs for a binary encoded two-output logic function that is obtained by converting the maximally asymmetric function f_1 in Table 1 as follows:*

$$X_1 = (x_1, x_2) \quad X_2 = (x_3, x_4) \quad X_3 = (x_5, x_6)$$
$$(0)_3 = (00)_2 \quad (1)_3 = (01)_2 \quad (2)_3 = (10)_2$$
$$f_1 = (f_{1h}, f_{1l}).$$

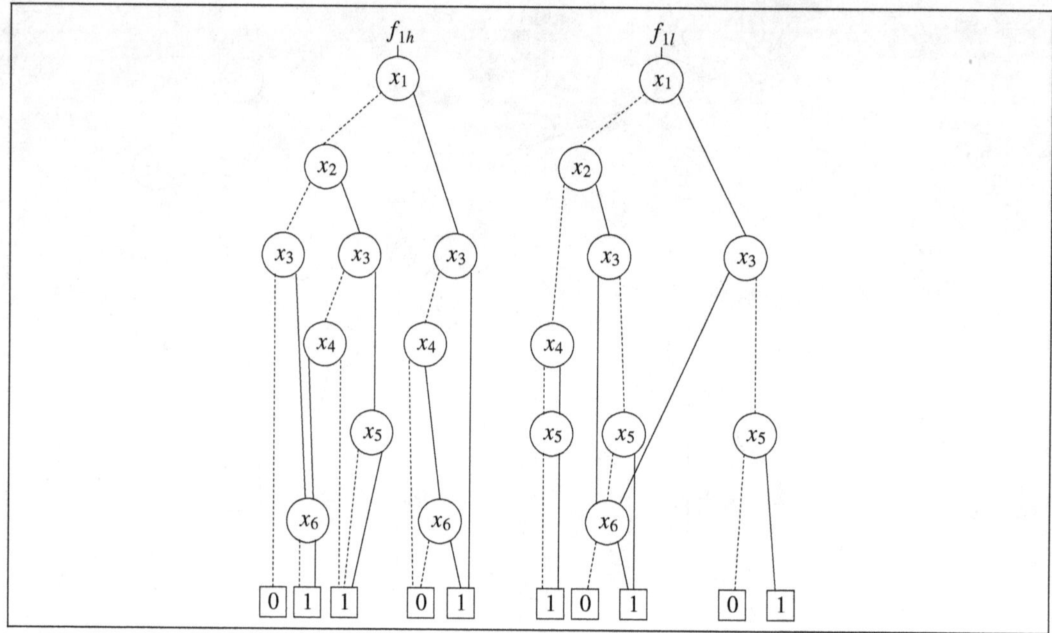

Figure 2: ZDDs for $f_1 = (f_{1h}, f_{1l})$

In these figures, dashed lines and solid lines denote 0*-edges and* 1*-edges, respectively. For simplicity, terminal nodes are not shared completely. In addition, each logic function is represented separately by a single-rooted decision diagram. However, multiple-output logic functions can be represented by multiple-rooted monolithic decision diagrams sharing equivalent sub-graphs among logic functions. Such multiple-rooted BDDs and ZDDs are called shared BDDs and shared ZDDs, respectively.* □

Definition 8. *A* ***multi-valued decision diagram (MDD)*** *[15] is an extension of a BDD to represent an* r*-valued function. It consists of* r *terminal nodes representing function values, and nonterminal nodes representing input variables. Each nonterminal node has* r *outgoing edges that correspond to the values of the input variables.*

Definition 9. *An* ***edge-valued MDD (EVMDD)*** *[22] is a variant of an MDD. It consists of one terminal node representing* 0 *and nonterminal nodes with edges having integer weights;* 0*-edges always have zero weights. In an EVMDD, the function value is represented as a sum of weights for edges traversed from the root node to the terminal node.*

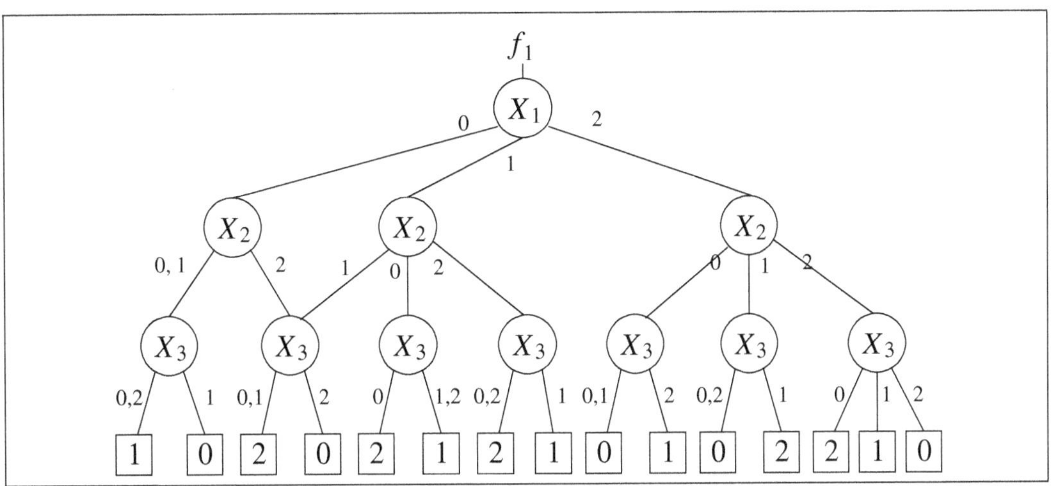

Figure 3: MDD for f_1

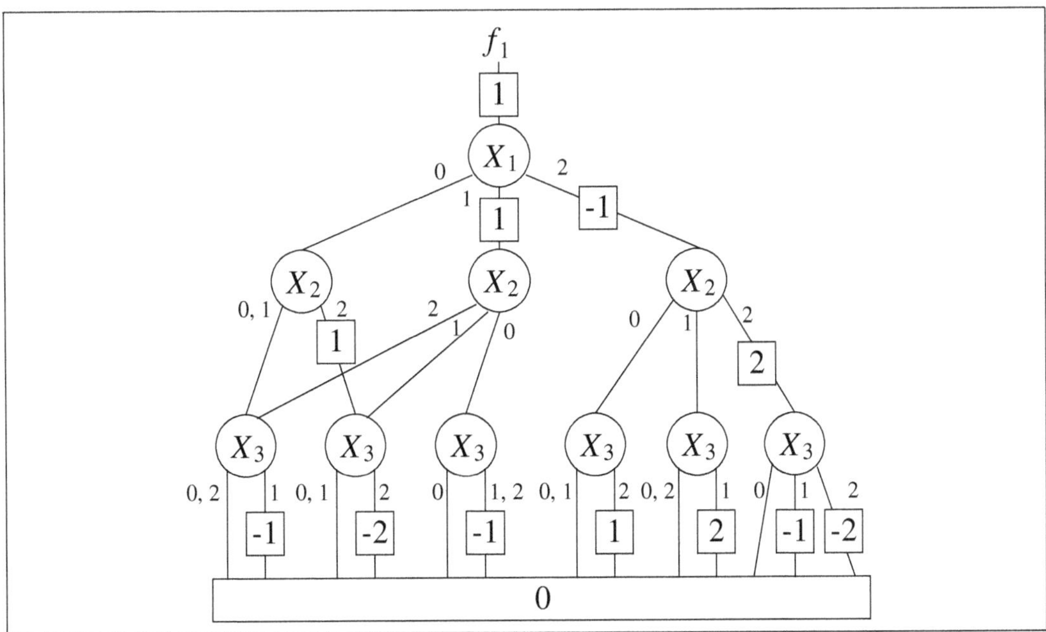

Figure 4: EVMDD for f_1

Example 3. *Figs. 3 and 4 show an MDD and an EVMDD for the maximally asymmetric function f_1, respectively. For simplicity, terminal nodes in the MDD are not shared completely. Note that in Fig. 4, the edge to the root node also has a weight* 1. □

3 Analysis of Maximal Asymmetries

This section begins by introducing the notation needed for characteristic analysis of maximally asymmetric functions. Then, we derive a characteristic of maximally asymmetric functions.

3.1 Notation for Characteristic Analysis

To analyze characteristics of maximally asymmetric functions, we introduce the notation used for symmetric functions [7, 8]. As shown in Example 1, function values of symmetric functions depend on *combinations of input values*. Thus, the set of input vectors can be classified into equivalence classes, each of which has the same combination of input values. A representative of each equivalence class is denoted as follows:

Definition 10. *For an n-variable r-valued function $f(X_1, X_2, \ldots, X_n)$, when input vectors $(X_1, X_2, \ldots, X_n)$ are classified into equivalence classes, each of which has the same combination of input values, a representative of an i-th equivalence class is*

$$\vec{\alpha}(i) = (\alpha_0(i), \alpha_1(i), \ldots, \alpha_{r-1}(i)),$$

*where $\alpha_j(i)$ denotes the number of variables whose values are j in the i-th equivalence class, and $\sum_{j=0}^{r-1} \alpha_j(i) = n$. In this paper, such equivalence classes of input vectors are called **α-equivalence classes**. Each equivalence class is denoted by $[\vec{\alpha}(i)]$, and $\#[\vec{\alpha}(i)]$ denotes the number of input vectors belonging to the class $[\vec{\alpha}(i)]$.*

Lemma 1. *For an n-variable r-valued function, the number of input vectors belonging to a $[\vec{\alpha}(i)]$ is*

$$\#[\vec{\alpha}(i)] = \frac{n!}{\alpha_0(i)!\alpha_1(i)!\ldots!\alpha_{r-1}(i)!}. \tag{1}$$

(Proof) This is the number of ways to distribute n distinct items into r distinct bins, which is the multinomial in (1). □

Lemma 2. *[8] For an n-variable r-valued function, the number of α-equivalence classes, N_α, is*

$$N_\alpha = \binom{n+r-1}{r-1}.$$

Since $N_\alpha < r^n$, symmetric functions are represented more compactly than their truth tables by using the notation of $\vec{\alpha}$ and its corresponding function value.

i	X_1	X_2	X_3	$\alpha_0(i)$	$\alpha_1(i)$	$\alpha_2(i)$	g_1
1	0	0	0	3	0	0	1
2	1	1	1	0	3	0	2
3	2	2	2	0	0	3	0
4	0 0 1	0 1 0	1 0 0	2	1	0	1
5	0 0 2	0 2 0	2 0 0	2	0	1	0
6	0 1 1	1 0 1	1 1 0	1	2	0	1
7	0 2 2	2 0 2	2 2 0	1	0	2	1
8	1 1 2	1 2 1	2 1 1	0	2	1	0
9	1 2 2	2 1 2	2 2 1	0	1	2	0
10	0 0 1 1 2 2	1 2 0 2 0 1	2 1 2 0 1 0	1	1	1	2

Table 2: Representation of g_1 using $\vec{\alpha}(i)$'s.

Example 4. *Table 2 shows the α-equivalence classes for g_1 in Table 1. Here, $N_\alpha = 10$ since $n = 3$ and $r = 3$.* □

In symmetric functions, all input vectors belonging to an α-equivalence class map to the same function value. Thus, we can decide whether a given function is symmetric or not by checking if all input vectors in an equivalence class map to only one function value. To implement this process, the following notation is introduced.

Definition 11. *In an n-variable r-valued function f, consider input vectors belonging to an α-equivalence class $[\vec{\alpha}(i)]$. Of $\#[\vec{\alpha}(i)]$ input vectors, the number of input vectors, X's, satisfying $f(X) = j$ is denoted by $v_j(f, i)$. Then, the distribution of input vectors to function values for the i-th equivalence class $[\vec{\alpha}(i)]$ is*

$$\vec{v}(f,i) = (v_0(f,i), v_1(f,i), \ldots, v_{r-1}(f,i)),$$

where values of $v_j(f,i)$'s are constrained by $\sum_{j=0}^{r-1} v_j(f,i) = \#[\vec{\alpha}(i)]$.

i	X_1	X_2	X_3	$\vec{\alpha}(i)$ $\alpha_0(i)$	$\alpha_1(i)$	$\alpha_2(i)$	g_1	$\vec{v}(g_1,i)$ $v_0(g_1,i)$	$v_1(g_1,i)$	$v_2(g_1,i)$	f_1	$\vec{v}(f_1,i)$ $v_0(f_1,i)$	$v_1(f_1,i)$	$v_2(f_1,i)$
1	0	0	0	3	0	0	1	0	1	0	1	0	1	0
2	1	1	1	0	3	0	2	0	0	1	2	0	0	1
3	2	2	2	0	0	3	0	1	0	0	0	1	0	0
4	0 0 1	0 1 0	1 0 0	2	1	0	1 1 1	0	3	0	0 1 2	1	1	1
5	0 0 2	0 2 0	2 0 0	2	0	1	0 0 0	3	0	0	1 2 0	1	1	1
6	0 1 1	1 0 1	1 1 0	1	2	0	1 1 1	0	3	0	0 1 2	1	1	1
7	0 2 2	2 0 2	2 2 0	1	0	2	1 1 1	0	3	0	0 1 2	1	1	1
8	1 1 2	1 2 1	2 1 1	0	2	1	0 0 0	3	0	0	0 1 2	1	1	1
9	1 2 2	2 1 2	2 2 1	0	1	2	0 0 0	3	0	0	2 0 1	1	1	1
10	0 0 1 1 2 2	1 2 0 2 0 1	2 1 2 0 1 0	1	1	1	2 2 2 2 2 2	0	0	6	1 2 1 2 0 0	2	2	2

Table 3: Representation of g_1 and f_1 using $\vec{\alpha}(i)$'s and $\vec{v}(f,i)$'s.

Example 5. *Table 3 shows distributions of input vectors to function values $\vec{v}(f,i)$'s for g_1 and f_1 shown in Table 1 with $\vec{\alpha}(i)$'s.* □

3.2 Computation Method for Asymmetry

As shown in Table 3, in each $\vec{v}(g,i)$ of symmetric functions g , only one element $v_j(g,i)$ has a non-zero value, and all other elements are zero. On the other hand, in asymmetric functions f, there exists a $\vec{v}(f,i)$ that has more than one non-zero $v_j(f,i)$. From this observation of distributions $\vec{v}(f,i)$'s, we have the following lemma:

Lemma 3. *A function g is symmetric iff for any α-equivalence class $[\vec{\alpha}(i)]$,*

$$\max(\vec{v}(g,i)) = \#[\vec{\alpha}(i)]$$

holds, where $\max(\vec{v}(g,i))$ *is the largest element $v_j(g,i)$ in $\vec{v}(g,i)$.*

(Proof) If g is symmetric, then from Definitions 3 and 10, it is clear that all the input vectors in an equivalence class $[\vec{\alpha}(i)]$ correspond to the same function value. Without loss of generality, let the function value be j. Then, from Definition 11, only $v_j(g,i)$ has the value of $\#[\vec{\alpha}(i)]$, and all others are zero. Thus, $\max(\vec{v}(g,i)) = \#[\vec{\alpha}(i)]$ holds.

If $\max(\vec{v}(g,i)) = \#[\vec{\alpha}(i)]$ holds, then from Definition 11, all $v_k(g,i)$'s other than the largest element $v_j(g,i)$ are zero since values of $v_k(g,i)$'s are constrained by $\sum_{k=0}^{r-1} v_k(f,i) = \#[\vec{\alpha}(i)]$. This means that all the input vectors in $[\vec{\alpha}(i)]$ correspond to the function value j. Thus, g is symmetric. □

From Lemma 3, when $\max(\vec{v}(f,i)) < \#[\vec{\alpha}(i)]$, a function f is asymmetric. For an asymmetric function f, its nearest symmetric function is obtained by the following lemma:

Lemma 4. *For an asymmetric function f, consider α-equivalence classes $[\vec{\alpha}(i)]$'s and distributions $\vec{v}(f,i)$'s of input vectors. For each $[\vec{\alpha}(i)]$, let $v_j(f,i) = \max(\vec{v}(f,i))$. Then, by assigning j as function values for input vectors in $[\vec{\alpha}(i)]$, we obtain the nearest symmetric function to f.*

(Proof) In the asymmetric function f, $v_j(f,i) = \max(\vec{v}(f,i))$ means that the most input vectors among ones in $[\vec{\alpha}(i)]$ correspond to the function value j. Since in symmetric functions, all the input vectors in $[\vec{\alpha}(i)]$ correspond to the same function value, by setting the function value to j, we have a symmetric function with the smallest Hamming distance from f. □

To compute the asymmetry of general functions f according to Definition 4, a symmetric function with the smallest Hamming distance from f has to be found. In general, such a search task tends to be intractable. However, we can easily find a symmetric function with the smallest Hamming distance by focusing on distributions of function values $\vec{v}(f,i)$ and by using Lemma 4. In addition, based on Lemma 4, we can efficiently compute the asymmetry of general functions f by the following theorem:

Theorem 1. *The asymmetry of an n-variable r-valued function f is*

$$\min_{g\in\Gamma} d(F,G) = \sum_{i=1}^{N_\alpha} (\#[\vec{\alpha}(i)] - \max(\vec{v}(f,i))), \tag{2}$$

where $\max(\vec{v}(f,i))$ is the largest element $v_j(f,i)$ in $\vec{v}(f,i)$.

(Proof) From Lemma 4, in the nearest symmetric function to f, all the input vectors in $[\vec{\alpha}(i)]$ correspond to the function value j. On the other hand, in f, $v_j(f,i)$ input vectors in $[\vec{\alpha}(i)]$ correspond to j. That is, other input vectors in $[\vec{\alpha}(i)]$ correspond to different function values than j. These input vectors contribute to the Hamming distance between F and G. Thus, by tallying the difference $(\#[\vec{\alpha}(i)] - \max(\vec{v}(f,i)))$, we can compute the asymmetry of f. □

Note that the right hand side of (2) computes the asymmetry of f without using the set of symmetric functions. Thus, Theorem 1 computes the asymmetry much more efficiently than Definition 4.

3.3 Characteristic of Maximally Asymmetric Functions

From Theorem 1, the maximum asymmetry occurs when all the $\max(\vec{v}(f,i))$'s are minimum. That is, when all the $\max(\vec{v}(f,i))$'s are minimum, their function f is maximally asymmetric. When the distribution of input vectors $\vec{v}(f,i)$ is **uniform**, $\max(\vec{v}(f,i))$ achieves the minimum. However, its converse does not hold. Even if a $\max(\vec{v}(f,i))$ is minimum, the distribution $\vec{v}(f,i)$ is not always uniform because the number of input vectors is not always an integer multiple of r positions specified by $\vec{v}(f,i)$. By considering discreteness of $\vec{v}(f,i)$, we have the following theorem on the minimum $\max(\vec{v}(f,i))$'s for maximally asymmetric functions f.

Theorem 2. *An n-variable r-valued function f is maximally asymmetric iff each $\vec{v}(f,i)$ for $[\vec{\alpha}(i)]$ satisfies the following:*

$$\max(\vec{v}(f,i)) = \left\lceil \frac{\#[\vec{\alpha}(i)]}{r} \right\rceil .$$

(Proof) From (2), the maximum asymmetry occurs when $\max(\vec{v}(f,i))$ is minimum across all i. Note that from Definition 11, a value of each element $v_j(f,i)$ in $\vec{v}(f,i)$ is constrained by $\sum_{j=0}^{r-1} v_j(f,i) = \#[\vec{\alpha}(i)]$. Each i contributes a part of the asymmetry to the right hand side of (2) a value that is independent of all other i. That is, each value, $\max(\vec{v}(f,i))$, should be minimum, independent of all other values. This occurs when

$$\max(\vec{v}(f,i)) = \left\lceil \frac{\#[\vec{\alpha}(i)]}{r} \right\rceil$$

for all i. □

If we follow only Definition 5, it is hard to decide if a given function f is maximally asymmetric by just examining function itself since we have to confirm that the asymmetry of f is the largest among all functions. However, by using Theorem 2, we can easily decide whether f is maximally asymmetric or not by just examining f itself.

Lemma 5. *For an n-variable r-valued maximally asymmetric function f, the smallest element in each $\vec{v}(f,i)$ is*

$$\min(\vec{v}(f,i)) = \begin{cases} 0 & \left(\frac{\#[\vec{\alpha}(i)]}{\max(\vec{v}(f,i))} \leq r-1\right) \\ \#[\vec{\alpha}(i)]\% \max(\vec{v}(f,i)) & \left(r-1 < \frac{\#[\vec{\alpha}(i)]}{\max(\vec{v}(f,i))} < r\right) \\ \max(\vec{v}(f,i)) & \left(\frac{\#[\vec{\alpha}(i)]}{\max(\vec{v}(f,i))} = r\right) \end{cases}$$

where % denotes the modulo operation.

(Proof) Each element in $\vec{v}(f,i)$ can be up to $\max(\vec{v}(f,i))$ while satisfying the necessary and sufficient condition for maximally asymmetric functions. If $\#[\vec{\alpha}(i)]$ is large enough that all the elements are $\max(\vec{v}(f,i))$, then the smallest element is $\max(\vec{v}(i))$. Otherwise, it is 0 or the remainder when dividing $\#[\vec{\alpha}(i)]$ by $\max(\vec{v}(f,i))$, depending on $\#[\vec{\alpha}(i)]$. □

Lemma 6. *Let $f_0, f_1, \ldots, f_{r-1}$ be $(n-1)$-variable r-valued maximally asymmetric functions. Consider an n-variable r-valued function composed of the r maximally asymmetric functions by the Shannon expansion. Then, there exists an n-variable function that is not maximally asymmetric.*

(Proof) Consider f_0, f_1, and f_2 in Table 4. These functions are maximally asymmetric. However, a 3-variable function f composed by the Shannon expansion:

$$f = X_3^0 f_0 + X_3^1 f_1 + X_3^2 f_2$$

is not maximally asymmetric as shown in Table 4, where X_3^i is a literal of X_3. The asymmetry of f is 10 that is smaller than the asymmetry, 16, of the MAF in Table 1 □

4 DD-Based Computation Methods for Asymmetry and MAFs

This section proposes decision diagram-based methods to compute asymmetry of functions and to generate benchmarks for maximally asymmetric functions.

4.1 Computation of Asymmetry on Decision Diagrams

Given a multiple-valued function represented by a decision diagram, its asymmetry is computed by the following two main processes:

X_1	X_2	X_3	f
0	0	0	2
1	1	1	0
2	2	2	0
0	0	1	1
0	1	0	1
1	0	0	0
0	0	2	1
0	2	0	0
2	0	0	1
0	1	1	2
1	0	1	1
1	1	0	2
0	2	2	2
2	0	2	0
2	2	0	0
1	1	2	2
1	2	1	1
2	1	1	2
1	2	2	0
2	1	2	2
2	2	1	2
0	1	2	2
0	2	1	1
1	0	2	0
1	2	0	0
2	0	1	2
2	1	0	1

X_1	X_2	f_0	f_1	f_2
0	0	2	1	1
1	1	2	0	2
2	2	0	2	0
0	1	1	2	2
1	0	0	1	0
0	2	0	1	2
2	0	1	2	0
1	2	0	1	0
2	1	1	2	2

Table 4: A function f composed of MAFs f_0, f_1, and f_2.

1. Traverse all paths in a decision diagram in the depth first manner, and make a table of $\vec{a}(i)$'s and $\vec{v}(i)$'s as in Table 3.
2. Compute the asymmetry using the table and Theorem 1.

Fig. 5 shows pseudo code for the Process 1. This procedure requires the root node of a decision diagram and the undefined input vector as its arguments. It traverses a path while setting elements of input vector one by one. When a terminal node

```
Make_Vtable (node, input_vector) {
  if (node is terminal) {
    Find equivalence classes [α⃗(i)]'s to which input_vector belongs;
    Update their v⃗(i)'s with the function value represented at the terminal;
    return;
  }

  for (i = 0; i < r; i++) {
    input_vector[node.variable] = i;
    Make_Vtable (node.edge[i], input_vector);
  }
  input_vector[node.variable] = Undefined;
  return;
}
```

Figure 5: Traversing a DD and making a table $\vec{\alpha}$ and $\vec{v}$.

is reached, equivalence classes $[\vec{\alpha}(i)]$'s to which the input vector belongs are found. Note that some elements in the input vector can remain undefined due to skipped nodes in a path of the decision diagram. Since all possible values are assigned to those elements, multiple classes $[\vec{\alpha}(i)]$'s are found. Then, their $\vec{v}(i)$'s are incremented by the function value represented at the terminal node.

4.2 Benchmark Generation of Maximally Asymmetric Functions

In applications where maximally asymmetric functions can be promising, benchmark functions are required to evaluate their usefulness. This subsection presents a method to randomly generate maximally asymmetric functions based on the characteristic derived in the previous section.

As shown in Theorem 2, in maximally asymmetric functions, $\max(\vec{v}(i))$'s in distributions $\vec{v}(i)$'s are minimized. Therefore, by generating function values randomly such that $\max(\vec{v}(i))$'s are minimized, we can generate a benchmark for maximally asymmetric function. Lemma 6 shows that from smaller maximally asymmetric functions, a larger maximally asymmetric function cannot be always composed by the Shannon expansion. Thus, we generate an n-variable maximally asymmetric function directly without using subfunctions. Algorithm 1 shows the algorithm to randomly generate a benchmark for a maximally asymmetric function in decision diagram form.

In Algorithm 1, steps 3 and 4 randomly generate a function value j in a range

Algorithm 1. *Overview of the generation algorithm*

Input: the numbers of variables n and values r
Output: a decision diagram DD_{fin} for an n-variable r-valued MAF
Initialize:
- *each $\vec{v}(i) = \vec{0}$, and*
- *construct DD_{fin} for the 0 constant function.*

1. *For each class $[\vec{\alpha}(i)]$, the following processes apply.*
   2. *For each input vector in $[\vec{\alpha}(i)]$, the following applies.*
      3. *Randomly generate a function value j satisfying $v_j(i) < \left\lceil \frac{\#[\vec{\alpha}(i)]}{r} \right\rceil$, where $v_j(i) \in \vec{v}(i)$.*
      4. *Increment $v_j(i)$.*
      5. *If $j = 0$, then skip the steps 6 and 7.*
      6. *Construct a decision diagram DD_{min} for an input vector and j (i.e., a minterm).*
      7. *Merge DD_{min} with DD_{fin}.*

of 0 to $r-1$ satisfying $v_j(i) \leq \left\lceil \frac{\#[\vec{\alpha}(i)]}{r} \right\rceil$. Step 6 constructs a decision diagram for a minterm, and then at step 7, it is merged with DD_{fin} by any of the following operations:

- *Max* operation for MDDs [15]
- *Add* operation for EVMDDs shown in Fig. 6
- *Logical OR* operation for BDDs [6]
- *Union* operation for ZDDs [20]

Since the DD_{fin} is initially set to the constant 0 function, we can skip the processes for constructing and merging decision diagrams if the generated function value j is 0. After all the iterations of steps 1 and 2, the decision diagram DD_{fin} representing an n-variable r-valued maximally asymmetric function is obtained.

Fig. 6 shows pseudo code for the Add operation for EVMDDs. For simplicity, pseudo code for the computed table is omitted in this figure. In Fig. 6, "height" denotes the height of EVMDD for f or g. This operation merges two EVMDDs in the depth first manner.

```
Add_EVMDD (f, g) {
  if (f = 0) return g;
  if (g = 0) return f;
  if (f.height > g.height) {
    for (i = 0; i < r; i++) {
      edge[i] = Add_EVMDD (f.edge[i], g);
      edge_val[i] = f.edge_val[i];
    }
    height = f.height;
  }
  else if (f.height < g.height) {
    for (i = 0; i < r; i++) {
      edge[i] = Add_EVMDD (f, g.edge[i]);
      edge_val[i] = g.edge_val[i];
    }
    height = g.height;
  }
  else {
    for (i = 0; i < r; i++) {
      edge[i] = Add_EVMDD (f.edge[i], g.edge[i]);
      edge_val[i] = f.edge_val[i] + g.edge_val[i];
    }
    height = f.height;
  }
  return get_node (height, edge, edge_val);
}
```

Figure 6: Add operation for EVMDDs f and g

5 Experimental Results

The proposed benchmark generation method is implemented on our own package for decision diagrams, and run on the following computer environment: CPU: Intel Core2 Quad Q6600 2.4GHz, memory: 4GB, OS: CentOS 5.7 Linux, and C-compiler: gcc -O2 (version 4.1.2). The variable order for decision diagrams is the natural order (i.e., $x_1, x_2, \ldots, x_n$, from the root node to a terminal node).

We randomly generated n-variable r-valued maximally asymmetric functions in MDDs, EVMDDs, BDDs, and ZDDs using the proposed method, for various n and r. All the decision diagrams represent the same functions. For each n and r,

n	r	Truth Tables (r^n words)	MDDs (# nodes)	Ratio (%)
3	3	27	14	52
4	3	81	33	40
5	3	243	66	27
6	3	729	148	20
7	3	2,187	390	18
8	3	6,561	1,106	17
9	3	19,683	3,187	16
10	3	59,049	8,865	15
11	3	177,147	22,275	13
12	3	531,441	48,237	9
13	3	1,594,323	108,254	7
3	4	64	25	38
4	4	256	82	32
5	4	1,024	250	24
6	4	4,096	592	14
7	4	16,384	1,621	10
8	4	65,536	5,717	9
9	4	262,144	22,101	8
10	4	1,048,576	87,636	8
3	5	125	36	29
4	5	625	157	25
5	5	3,125	727	23
6	5	15,625	2,749	18
7	5	78,125	7,009	9
8	5	390,625	22,656	6
9	5	1,953,125	100,781	5

Table 5: Comparison in size of truth tables and MDDs.

we generated 10 maximally asymmetric functions. Other details are shown in the following subsections.

5.1 Comparison with Truth Tables

We begin by comparing the size of MDDs with the size of truth tables. For maximally asymmetric functions, suitable representations have not been well-known yet. Thus, we use the size of truth tables as a baseline for size evaluation of decision diagrams.

Table 5 compares the size of MDDs with the size of truth tables for n-variable r-valued maximally asymmetric functions. The size of the truth table is r^n, the total number of input vectors. The size of MDDs is the average number of nodes in

10 MDDs for the randomly generated maximally asymmetric functions. In Table 5, the average is rounded to the nearest integer. The column "Ratio" shows the ratio of the size of MDD to the size of truth table.

From Table 5, we can see that the size of MDDs is much smaller than that of truth tables. Thus, providing benchmarks for maximally asymmetric functions in MDD form is more useful than providing them in truth table form.

5.2 Comparison with Random Functions

To show the difference between maximally asymmetric functions and random functions, we compared them. We define random functions as follows:

Definition 12. *A function of which function values are generated only by uniform random numbers is called a* ***random function****.*

Table 6 compares maximally asymmetric functions with randomly generated functions, in terms of the number of nodes in an MDD and asymmetry. We randomly generated 10 n-variable r-valued functions by generating functions values using uniform random numbers. Table 6 shows the average number of nodes in an MDD and the average asymmetry. The columns "MAF" and "Random" show the results for maximally asymmetric functions and randomly generated functions, respectively. The column "UB of nodes in MDDs" shows the upper bound on the number of nodes in an MDD for an n-variable r-valued function. It is obtained by

$$\frac{r^{n-l}-1}{r-1} + r^{r^l},$$

where l is the largest integer satisfying $n - l \geq r^l$ [23].

From Table 6, we can see that the size of MDDs for maximally asymmetric functions is almost equal to the size of MDDs for randomly generated functions. In addition, their size is almost equal to its upper bound. Thus, the maximally asymmetric functions belong to the worst class, in which functions require large MDDs, as randomly generated functions.

However, randomly generated functions have smaller asymmetry than maximally asymmetric functions. In fact, no maximally asymmetric functions were obtained by just generating function values using uniform random numbers. From this observation, we can conclude that it is hard to generate maximally asymmetric functions unless the proposed method is used.

		# nodes		UB of	Asymmetry	
		MAF	Random	# nodes in	MAF	Random
n	r			MDDs [23]		
3	3	14	14	16	16	9
4	3	33	31	40	48	37
5	3	66	66	67	153	127
6	3	**148**	**148**	**148**	483	418
7	3	390	388	391	1,449	1,335
8	3	1,106	1,105	1,120	4,356	4,148
9	3	3,187	3,186	3,307	13,120	12,671
10	3	8,865	8,872	9,868	39,360	38,560
11	3	22,275	22,299	29,524	118,089	116,707
12	3	48,237	48,234	49,207	354,288	351,822
13	3	108,254	108,253	108,256	1,062,864	1,058,003
3	4	**25**	24	**25**	40	28
4	4	82	80	89	186	144
5	4	250	250	341	744	649
6	4	592	593	597	3,052	2,796
7	4	**1,621**	**1,621**	**1,621**	12,236	11,649
8	4	**5,717**	**5,717**	**5,717**	49,146	47,741
9	4	**22,101**	**22,101**	**22,101**	196,584	193,396
10	4	87,636	**87,637**	**87,637**	786,396	779,458
3	5	**36**	**36**	**36**	80	62
4	5	157	159	161	465	390
5	5	727	727	786	2,496	2,190
6	5	2,749	2,750	3,906	12,480	11,642
7	5	7,009	7,012	7,031	62,450	60,148
8	5	**22,656**	**22,656**	**22,656**	312,400	306,376
9	5	**100,781**	**100,781**	**100,781**	1,562,325	1,546,763

Boldfaced numbers show that the number of nodes in an MDD reaches its upper bound.

Table 6: Comparison of MAFs with random functions.

5.3 Comparison of Decision Diagrams

To investigate a decision diagram appropriate for maximally asymmetric functions, we compare sizes of MDDs, EVMDDs, BDDs, and ZDDs for them. To represent maximally asymmetric functions by BDDs and ZDDs, we convert the multiple-valued functions into multiple-output binary logic functions by using two kinds of encoding methods: the natural binary encoding and the one-hot encoding. These encoding can produce *don't cares* in function values. In this experiment, 0 is assigned to the

				Binary			One-hot	
n	r	MDDs	EVMDDs	BDDs	ZDDs	ZDDs for CFs	BDDs	ZDDs for CFs
3	3	14	12	28	20	25	65	33
4	3	33	28	67	49	57	151	73
5	3	66	59	161	117	118	364	153
6	3	148	140	400	297	277	899	347
7	3	390	382	977	730	732	2,149	884
8	3	1,106	1,097	2,354	1,808	1,948	5,100	2,342
9	3	3,187	3,172	5,953	4,752	4,990	12,780	6,100
10	3	8,865	8,822	16,013	13,205	12,895	34,423	16,087
11	3	22,275	22,069	45,143	37,721	32,874	97,235	–
12	3	48,237	47,771	128,003	106,839	78,528	277,474	–
13	3	108,254	107,742	358,213	296,072	197,590	–	–
3	4	25	21	43	42	50	152	71
4	4	82	73	131	130	141	446	208
5	4	250	213	421	422	439	1,446	586
6	4	592	513	1,270	1,271	1,293	4,898	1,611
7	4	1,621	1,540	4,312	4,312	4,341	16,237	5,026
8	4	5,717	5,636	16,116	16,115	16,179	51,805	18,229
9	4	22,101	22,020	58,677	58,673	58,848	162,717	–
10	4	87,636	87,555	187,742	187,746	188,031	527,256	–
3	5	36	32	129	89	101	287	129
4	5	157	150	446	335	388	1,049	501
5	5	727	682	1,776	1,459	1,623	4,349	1,908
6	5	2,749	2,310	7,367	6,053	5,742	18,973	6,595
7	5	7,009	5,997	32,802	26,558	19,381	82,146	22,267
8	5	22,656	21,632	153,247	123,700	81,852	344,917	–
9	5	100,781	99,757	704,325	567,252	393,172	–	–

"–": DDs could not be produced due to memory overflow or too long computation time.

Table 7: Comparison of MDDs, EVMDDs, BDDs, and ZDDs.

don't cares. The multiple-output logic functions are represented by shared BDDs and shared ZDDs. In addition, by using characteristic functions, the multiple-output logic functions are converted into single-output logic functions, and they are represented by single ZDDs.

Table 7 compares MDDs, EVMDDs, BDDs, and ZDDs for maximally asymmetric functions, in terms of the number of nodes. Similarly to the experiment for Table 5, 10 maximally asymmetric functions are generated, and the average number of nodes in their decision diagrams are shown in Table 7. The columns "BDDs" and "ZDDs" show the numbers of nodes in shared BDDs and shared ZDDs for the binary encoded functions, respectively. The column "ZDDs for CFs" shows the number of nodes in

ZDDs for characteristic functions of the encoded multiple-output functions.

As shown in Table 7, EVMDDs are the smallest among the decision diagrams used in this experiment. However, there is little difference between the sizes of MDDs and EVMDDs. The binary encoding and the one-hot encoding increase the number of input variables, resulting in much larger BDDs and ZDDs. There are some cases where by using characteristic functions, the size of ZDDs is reduced significantly. However, for maximally asymmetric functions, ZDDs for CFs are much larger than MDDs and EVMDDs. Thus, among these decision diagrams, MDDs and EVMDDs are suitable for maximally asymmetric functions.

6 Conclusion and Future Works

This paper derives a new characteristic of maximally asymmetric function, and proposes an automatic generation method of maximally asymmetric functions in decision diagram form. The derived characteristic is based on the computation method, and the proposed benchmark generation method uses the characteristic. The above things are the main contributions of this paper.

The proposed method efficiently generates rare maximally asymmetric functions, and thus, it will be helpful to find an application where the use of maximally asymmetric functions can be promising. The generated benchmark functions and their generator could be provided. This paper also shows sizes of MDDs, EVMDDs, BDDs, and ZDDs for maximally asymmetric functions. Among these decision diagrams, MDDs and EVMDDs are suitable for representation of maximally asymmetric functions. However, their sizes are almost equal to the upper bounds, and thus, the maximally asymmetric functions belong to the same worst class as random functions in terms of decision diagram size.

Since time complexity of the proposed generation method is $O(r^n)$, a method with less complexity would be more useful. Our future work includes reduction of time complexity and proposal of more scalable methods. In addition, investigating a more compact representation method would be interesting and helpful in various applications.

Acknowledgments

This research is partly supported by the JSPS KAKENHI Grant (C), No.23K11038, 2023. The reviewers' comments were helpful in improving the paper.

References

[1] S. B. Akers, "Binary decision diagrams," *IEEE Trans. Comput.*, Vol. C-27, No. 6, pp. 509–516, 1978.

[2] F. Armknecht, C. Carlet, P. Gaborit, S. Künzli, W. Meier, and O. Ruatta, "Efficient computation of algebraic immunity for algebraic and fast algebraic attacks," Adv. in Crypt. - EUROCRYPT 2006, LNCS, Vol. 4004, pp. 147–164, Springer-Verlag, 2006.

[3] A. Bogdonov and A. Rosen, "Pseudorandom functions: three decades later," in Y. Lindel (ed.) *Tutorials on the Found. of Crypto.*, pp. 79-158, Springer Inter. Publ. AG. Part of Springer Nature, 2017.

[4] R. C. Born, "An iterative technique for determining the minimal number of variables for a totally symmetric function with repeated variables," *IEEE Trans. Comput.*, Vol. C-21, No. 10, pp. 1129–1131, 1972.

[5] S. Boztas, R. Hammons, and P. V. Kumar, "4-phase sequences with near optimum correlation properties," *IEEE Trans. Infor. Theory*, Vol. 40, pp. 1101-1113, 1992.

[6] R. E. Bryant, "Graph-based algorithms for Boolean function manipulation," *IEEE Trans. Comput.*, Vol. C-35, No. 8, pp. 677–691, 1986.

[7] J. T. Butler and T. Sasao, "On the properties of multiple-valued functions that are symmetric in both variable values and labels," *Proc. of 28th International Symposium on Multiple-Valued Logic*, pp. 83-88, 1998.

[8] J. T. Butler and T. Sasao, "Maximally asymmetric multiple-valued functions," *Proc. of 49th International Symposium on Multiple-Valued Logic*, pp. 188-193, 2019.

[9] J. T. Butler and T. Sasao, "Enumerative analysis of asymmetric functions," *Proc. of Reed-Muller Workshop 2019*, pp. 3-11, 2019.

[10] J. T. Butler and T. Sasao, "Properties of multiple-valued partition functions," *Proc. of 50th International Symposium on Multiple-Valued Logic*, pp. 82-87, 2020.

[11] A. Canteaut and M. Videau, "Symmetric Boolean functions," *IEEE Trans. Infor. Theory*, Vol. 51, No. 8, pp. 2791–2811, 2005.

[12] B. Dahlberg, "On symmetric functions with redundant variables - weighted functions," *IEEE Trans. Comput.*, Vol. C-22, No. 5, pp. 450–458, 1973.

[13] O. Goldreich, S. Goldwasser, and S. Micali, "On the cryptographic application of random functions," in G. R. Blakley and D. Chaum (Eds.) *Advances in Cryptology - CRYPTO'84*, LNCS 196, pp. 276-288, 1985.

[14] O. Goldreich, S. Goldwasser, and S. Micali, "How to construct random functions," *Journal of the Association for Computing Machinery*, Vol. 33, No. 4, pp. 792-807, 1986.

[15] T. Kam, T. Villa, R. K. Brayton, and A. L. Sangiovanni-Vincentelli, "Multi-valued decision diagrams: Theory and applications," *Multiple-Valued Logic: An International Journal*, Vol. 4, No. 1-2, pp. 9–62, 1998.

[16] C. Karanikas, N. Atreas, and R. S. Stankovic, "Bent functions, bent permutations and a variety of methods to construct them," *Proc. of Reed-Muller Workshop 2019*, pp. 12-17, 2019.

[17] Z. Kohavi, *Switching and Finite Automata Theory*, McGraw-Hill Book Company, 1979.

[18] D. T. Lee and S. J. Hong, "An algorithm for transformation of an arbitrary switching function to a completely symmetric function," *IEEE Trans. Comput.*, Vol. C-25, No. 11, pp. 1117–1123, 1976.

[19] S. Maitra and P. Sarkar, "Maximum nonlinearity of symmetric Boolean functions on odd number of variables," *IEEE Trans. Infor. Theory*, Vol. 48, No. 9, pp. 2626–2630, 2002.

[20] S. Minato, "Zero-suppressed BDDs for set manipulation in combinatorial problems," *Proc. of 30th Design Automation Conference*, pp. 272–277, 1993.

[21] C. Moraga, M. Stankovic, and R. S. Stankovic, "On ternary symmetric bent functions," *Proc. of 50th International Symposium on Multiple-Valued Logic*, pp. 76–81, 2022.

[22] S. Nagayama, T. Sasao, and J. T. Butler, "A systematic design method for two-variable numeric function generators using multiple-valued decision diagrams," *IEICE Trans. on Information and Systems*, Vol. E93-D, No. 8, pp. 2059–2067, 2010.

[23] S. Nagayama, T. Sasao, and J. T. Butler, "Analysis of multi-state systems with multi-state components using EVMDDs," *Proc. of 42nd International Symposium on Multiple-Valued Logic*, pp.122-127, 2012.

[24] S. Nagayama, T. Sasao, and J. T. Butler, "On decision diagrams for maximally asymmetric functions," *Proc. of 52nd International Symposium on Multiple-Valued Logic*, pp. 164–169, 2022.

[25] J. Pieprzyk and C. X. Qu, "Fast hashing and rotation symmetric functions," *Journal of Universal Computer Science*, Vol. 5, No. 1, pp. 20–31, 1999.

[26] O. S. Rothaus, "On "bent" functions," *Journal of Combinatorial Theory, Series A*, Vol. 20, Issue 3, pp. 300-305, 1976.

[27] P. Sarkar and S. Maitra, "Balancedness and correlation immunity of symmetric Boolean functions," *Proc. of R. C. Bose Centenary Symp.*, Vol. 15, pp. 176–181, 2003.

[28] P. Savicky, "On the bent Boolean functions that are symmetric," *European J. Combinatorics*, Vol. 15, Issue 4, pp. 407-410, 1994.

[29] J. L. Shafer, S. Schneider, J. T. Butler, and P. Stanica, "Enumeration of bent Boolean functions by reconfigurable computer," *Proc. of 18th International Symposium on Field-Programmable Custom Computing Machines*, pp. 265-272, 2010.

[30] P. Stanica, T. Sasao, and J. T. Butler, "Distance duality on some classes of Boolean functions," *Journal of Combinatorial Mathematics and Combinatorial Computing*, Vol. 107, pp. 181-198, 2018.

[31] R. Stanley, *Enumerative Combinatorics, Vol. 1*, 2nd edition, Cambridge University Press, 2012.

[32] N. Tokareva, *Bent functions: Results and Applications to Cryptography*, Academic Press, 2015.

[33] K. Yang, Y.-K. Kim, and P. V. Kumar, "Quasi-orthogonal sequences for code-division multiple-access systems," *IEEE Trans. Infor. Theory*, Vol. 46, pp. 982-993, 2000.

[34] S. S. Yau and Y. S. Tang, "Transformation of an arbitrary switching function to a

totally symmetric function," *IEEE Trans. Comput.*, Vol. C-20, No. 12, pp. 1606–1609, Dec. 1971.

Received

p-valued Maiorana-McFarland Functions Structure of Their Reed-Muller Spectra

Claudio Moraga
Technical University of Dortmund, 44221 Dortmund, Germany
claudio.moraga@tu-dortmund.de

Radomir Stanković
Mathematical Institute of SASA, 11000 Belgrade, Serbia
radomir.stankovic@gmail.com

Milena Stanković
University of Niš, Faculty of Electronic Engineering, 18 000 Niš, Serbia
milena.stankovic@elfak.ni.ac.rs

Abstract

p-valued Maiorana-McFarland bent functions exist, as in the binary case, only for an even number of variables and they are normally associated to their Discrete Fourier spectra in order to specify their bentness through the flatness of the spectra. In this paper, a closely related, but different approach is considered. Structural properties of the Reed-Muller spectra of p-valued Maiorana-McFarland bent functions are studied when $p > 2$ is a prime. It is shown that the Reed-Muller spectra of p-valued Maiorana-McFarland bent functions have a regressive structure, parameterized by p and k, where $n = 2k$ denotes the even number of variables.

1 Introduction

Bent functions were introduced by Oscar Rothaus in 1976 [17] as the most non-linear Boolean functions. This property attracted the interest of researchers in the areas of Coding Theory and Cryptography. In [17], Rothaus provided a simple, but effective, method to generate Boolean bent functions. This method was strongly improved,

The authors gladly acknowledge the strong support given by the Reviewers of a preliminary version of this paper with their constructive criticism and suggestions for further research.

independently, by James Maiorana [6] and R. McFarland [7], which gave origin to the "Maiorana-McFarland (MMF) class of bent functions."

Bent functions were extended to the multiple-valued domain about a decade later by P. V. Kumar et al. [4] and in the spectral domain, by M. Luis and C. Moraga [5]. p-valued Maiorana-McFarland bent functions were possibly introduced in [9] and [10] and continued to be studied up to [13]. See, however, the books [8] and [20] for further references. Moreover, it should be recalled, that the Discrete Fourier transform converges to the Walsh transform [22] when working with Boolean functions, and to the Vilenkin-Chrestenson transform [21], [1] when working in the p-valued domain.

In the Boolean domain, the Reed-Muller transform [16], [15] is another view the Zhegalkin polynomials [24], [25]. The extension to the prime multiple-valued domain was introduced by Green and Taylor [2] in 1976. In the present paper, the Reed-Muller transform matrix is obtained as the inverse of the Reed-Muller basis in the ring of non-negative integers modulo p prime [11], [18].

The characterization of the Maiorana-McFarland Boolean bent functions (i.e., $p = 2$) based on the Reed-Muller transform was shown in [14] and it may be considered as a basic step and motivation for the study reported in the present chapter. Results obtained for $p > 2$ are consistent with the results obtained with $p = 2$.

2 Formalisms

Notation:

Let p be a prime, with $p > 2$. $Y_p(k)$ will denote a p-valued square matrix or a vector of dimension p^k. From the context it will be clear whether it represents a matrix or a vector. Moreover, if the valuedness is known, the index p may be omitted.

Definition 1. *Let Q stand for a vector or a square matrix. Then $\mathbf{1}^Q$ denotes a vector or a square matrix of the same dimension as Q, with all entries equal to* 1. *Similarly with $\mathbf{0}^Q$ and other values of the value set. Moreover, $\mathbf{1}^q$ denotes a vector or a square matrix of the same dimension as Q, with its first entry equal to* 1 *and all other entries equal to* 0.

Definition 2. *Let A be a matrix. Then $vec(A)$ represents the vectorizing operation, which concatenates the columns of A to build a column vector [3].*

Since in this paper, unless other specified, matrices will be square, the length of $vec(A)$ equals the square of the length of a side of A. Moreover, the inverse operation vec^{-1} is uniquely defined. To simplify the notation, ω will denote vec^{-1}. Thus, $\omega(vec(A)) = A$.

Definition 3. *Let A and B be matrices of the same dimensions. These matrices will be called disjoint if for all rows, whenever a row of a matrix has an entry larger than 0 then the corresponding row of the other matrix is a 0-row. Clearly, both matrices may share common 0-rows.*

The following Lemmas will support the below disclosed calculations.

Lemma 1. *(Adapted from Lemma 4.3.1 of [3])*
Let A, X, and B be square matrices of the same dimensions. Then:

$$(A \otimes A)vec(X) = vec(B) \leftrightarrow A \cdot X \cdot A^T = B. \tag{1}$$

Lemma 2. *[14]*
Let $Y(k)$ be a square matrix and G be a row vector of length p^k. Then:

$$Y(k)(\mathbf{1}^G \otimes G) = \langle Y(k) \cdot \mathbf{1}^G \rangle \otimes G. \tag{2}$$

In analogy to the Maiorana-McFarland method to generate Boolean bent functions, the following equation from [9], [10] will be used:

$$F = vec\langle M^{[k]} \cdot P(k) \oplus (\mathbf{1}^G \otimes G) \rangle, \tag{3}$$

where F denotes the value vector of an n-place MMF p-valued function, $n = 2k$, M is a $(p \times p)$ matrix, whose entries represent the value table of the product of two variables, $M^{[k]}$ indicates the k-fold tensor sum of M with itself [9], $P(k)$ represents a $(p^k \times p^k)$ permutation matrix, G is the row value vector of an arbitrary k-place p-valued function. In what follows, we restrict G to be other than the constant 0 to avoid a strong restriction of F as may be seen in Eqs. (3) (4). Further details are shown after Eq. (10).

From (3) follows:

$$\omega F = M^{[k]} \cdot P(k) \oplus (\mathbf{1}^G \otimes G). \tag{4}$$

3 Reed-Muller spectrum of ternary MMF functions

In this section, we study the RM spectrum of ternary MMF functions for different values of the number of variables n.

3.1 Case $k = 1$, $n = 2$, $3^k = 3$.

The basic Reed-Muller Basis is denoted symbolically $X_3(1) = \begin{bmatrix} 1 & x & x^2 \end{bmatrix}$ and in $(\mathbb{Z}_3, +, \cdot)$ is

$$X_3(1) = \begin{bmatrix} 1 & 0 & 0 \\ 1 & 1 & 1 \\ 1 & 2 & 1 \end{bmatrix}.$$

The basic Reed-Muller transform matrix is obtained as the inverse of the Basis [10], [11], [18], [19].

$$RM_3(1) = \begin{bmatrix} 1 & 0 & 0 \\ 0 & 2 & 1 \\ 2 & 2 & 2 \end{bmatrix}. \tag{5}$$

Furthermore, the Reed-Muller transform matrix has a Kronecker product structure [18], i.e.,

$$RM_3(2) = RM_3(1) \otimes RM_3(1) \bmod 3. \tag{6}$$

Notice that the bottom row of an RM matrix is a constant row and, except for the first row, all other rows add up to 0 mod 3.

Additionally, $M = \begin{bmatrix} 0 & 0 & 0 \\ 0 & 1 & 2 \\ 0 & 2 & 1 \end{bmatrix}$. Clearly, the first row of M is a 0-row and a permutation of its columns will obviously preserve this property. Moreover, the sum of the entries of every row and column of M is congruent with 0 mod 3.

With (6) and (3) the RM-spectrum of an MMF function on n variables is given by:

$$RM_3(2) \cdot F = RM_3(2) \cdot \langle vec\langle M \cdot P(1) \oplus (\mathbf{1}^G \otimes G)\rangle\rangle. \tag{7}$$

and with Lemma 1

$$\begin{aligned} RM_3(2) \cdot F &= RM_3(1) \cdot \omega F \cdot RM_3(1)^T \\ &= RM_3(1) \cdot \langle M \cdot P(1) \oplus (\mathbf{1}^G \otimes G)\rangle \cdot RM_3(1)^T \\ RM_3(2) \cdot F &= RM_3(1) \cdot M \cdot P(1) \cdot RM_3(1)^T \\ &\quad \oplus RM3(1) \cdot (\mathbf{1}^G \otimes G) \cdot RM_3(1)^T. \end{aligned} \tag{8}$$

Let

$$\begin{aligned} Alpha_3(1) &= RM_3(1) \cdot M \cdot P(1) \cdot RM_3(1)^T, \\ Beta_3(1) &= RM_3(1) \cdot (\mathbf{1}^G \otimes G) \cdot RM_3(1)^T. \end{aligned} \tag{9}$$

Recall that the first row of $M \cdot P(1)$ is a 0-row; then the first row of Alpha will also be a 0-row.

With Lemma 2,

$$\begin{aligned} RM_3(1) \cdot (\mathbf{1}^G \otimes G) &= (RM_3(1) \cdot \mathbf{1}^G) \otimes G \\ &= \left(\begin{bmatrix} 1 & 0 & 0 \\ 0 & 2 & 1 \\ 2 & 2 & 2 \end{bmatrix} \cdot \begin{bmatrix} 1 \\ 1 \\ 1 \end{bmatrix} \right) \otimes G \\ &= \begin{bmatrix} 1 \\ 0 \\ 0 \end{bmatrix} \otimes G = \begin{bmatrix} g_1 & g_2 & g_3 \\ 0 & 0 & 0 \\ 0 & 0 & 0 \end{bmatrix}, \end{aligned}$$

therefore,

$$Beta_3(1) = \begin{bmatrix} g_1 & g_2 & g_3 \\ 0 & 0 & 0 \\ 0 & 0 & 0 \end{bmatrix} \cdot \begin{bmatrix} 1 & 0 & 2 \\ 0 & 2 & 2 \\ 0 & 1 & 2 \end{bmatrix} = \begin{bmatrix} g_1 & 2g_2 + g_2 & 2(g_1 + g_2 + g_3) \\ 0 & 0 & 0 \\ 0 & 0 & 0 \end{bmatrix} . \tag{10}$$

It becomes apparent that the first row of Beta is the only non-0 row. Therefore, *Alpha* and *Beta* are disjoint matrices. (If G had been chosen to be the constant 0 function, then *Beta* would become a 0-matrix, obviously also disjoint with *Alpha*.) Furthermore, it may be shown that the steps leading to equation (10) may be extended to other values of p prime.

Notice that

$$(Beta_3(1))^T = RM_3(1) \cdot \left[\begin{array}{ccc} G^T & \mathbf{0}^G & \mathbf{0}^G \end{array} \right] = \left[\begin{array}{ccc} (RM_3(1) \cdot G^T) & \mathbf{0}^G & \mathbf{0}^G \end{array} \right],$$

therefore, the first row of *Beta* equals $(RM_3(1) \cdot G^T)^T$. An extension to larger values of k is straightforward.

Example 1. *In (3), let* $P(1) = \begin{bmatrix} 0 & 0 & 1 \\ 1 & 0 & 0 \\ 0 & 1 & 0 \end{bmatrix}$ *and* $G = \begin{bmatrix} 0 & 2 & 1 \end{bmatrix}$*. Then,*

$$\begin{aligned} Alpha_3(1) &= RM_3(1) \cdot M \cdot P(1) \cdot RM_3(1)^T \\ &= \begin{bmatrix} 1 & 0 & 0 \\ 0 & 2 & 1 \\ 2 & 2 & 2 \end{bmatrix} \cdot \begin{bmatrix} 0 & 0 & 0 \\ 0 & 1 & 2 \\ 0 & 2 & 1 \end{bmatrix} \cdot \begin{bmatrix} 0 & 0 & 1 \\ 1 & 0 & 0 \\ 0 & 1 & 0 \end{bmatrix} \cdot \begin{bmatrix} 1 & 0 & 2 \\ 0 & 2 & 2 \\ 0 & 1 & 2 \end{bmatrix} \\ &= \begin{bmatrix} 0 & 0 & 0 \\ 1 & 1 & 0 \\ 0 & 0 & 0 \end{bmatrix}. \end{aligned}$$

In $Beta_3(1)$*, the first row is*

$$row\ 1 = (RM_3(1) \cdot G^T)^T = \left(\begin{bmatrix} 1 & 0 & 0 \\ 0 & 2 & 1 \\ 2 & 2 & 2 \end{bmatrix} \cdot \begin{bmatrix} 0 \\ 2 \\ 1 \end{bmatrix} \right)^T = \begin{bmatrix} 0 & 2 & 0 \end{bmatrix}.$$

3.2 Case $k = 2$, $n = 4$, $p^k = 9$

With (6),

$$RM(4) = RM(2) \otimes RM(2) \bmod 3 = \langle RM(1) \rangle^{\otimes 4} \bmod 3.$$

and by using results from [3],

$$\begin{aligned} M^{[2]} &= (M \otimes \mathbf{1}^M) \oplus (\mathbf{1}^M \otimes M) \\ &= \begin{bmatrix} 0 & 0 & 0 & 0 & 0 & 0 & 0 & 0 & 0 \\ 0 & 1 & 2 & 0 & 1 & 2 & 0 & 1 & 2 \\ 0 & 2 & 1 & 0 & 2 & 1 & 0 & 2 & 1 \\ 0 & 0 & 0 & 1 & 1 & 1 & 2 & 2 & 2 \\ 0 & 1 & 2 & 1 & 2 & 0 & 2 & 0 & 1 \\ 0 & 2 & 1 & 1 & 0 & 2 & 2 & 1 & 0 \\ 0 & 0 & 0 & 2 & 2 & 2 & 1 & 1 & 1 \\ 0 & 1 & 2 & 2 & 0 & 1 & 1 & 2 & 0 \\ 0 & 2 & 1 & 2 & 1 & 0 & 1 & 0 & 2 \end{bmatrix}. \end{aligned} \tag{11}$$

Example 2. *Let*

$$P(2) = \begin{bmatrix} 0 & 0 & 1 \\ 1 & 0 & 0 \\ 0 & 1 & 0 \end{bmatrix} \otimes \begin{bmatrix} 0 & 1 & 0 \\ 1 & 0 & 0 \\ 0 & 0 & 1 \end{bmatrix} = \begin{bmatrix} 0 & 0 & 0 & 0 & 0 & 0 & 0 & 1 & 0 \\ 0 & 0 & 0 & 0 & 0 & 0 & 1 & 0 & 0 \\ 0 & 0 & 0 & 0 & 0 & 0 & 0 & 0 & 1 \\ 0 & 1 & 0 & 0 & 0 & 0 & 0 & 0 & 0 \\ 1 & 0 & 0 & 0 & 0 & 0 & 0 & 0 & 0 \\ 0 & 0 & 1 & 0 & 0 & 0 & 0 & 0 & 0 \\ 0 & 0 & 0 & 0 & 1 & 0 & 0 & 0 & 0 \\ 0 & 0 & 0 & 1 & 0 & 0 & 0 & 0 & 0 \\ 0 & 0 & 0 & 0 & 0 & 1 & 0 & 0 & 0 \end{bmatrix}$$

and $G = \begin{bmatrix} 0 & 2 & 1 \end{bmatrix} \otimes \begin{bmatrix} 1 & 0 & 2 \end{bmatrix} \mod 3$.

Then,

$$\begin{aligned} M^{[2]} \cdot P(2) &= \begin{bmatrix} 0 & 0 & 0 & 0 & 0 & 0 & 0 & 0 & 0 \\ 0 & 1 & 2 & 0 & 1 & 2 & 0 & 1 & 2 \\ 0 & 2 & 1 & 0 & 2 & 1 & 0 & 2 & 1 \\ 0 & 0 & 0 & 1 & 1 & 1 & 2 & 2 & 2 \\ 0 & 1 & 2 & 1 & 2 & 0 & 2 & 0 & 1 \\ 0 & 2 & 1 & 1 & 0 & 2 & 2 & 1 & 0 \\ 0 & 0 & 0 & 2 & 2 & 2 & 1 & 1 & 1 \\ 0 & 1 & 2 & 2 & 0 & 1 & 1 & 2 & 0 \\ 0 & 2 & 1 & 2 & 1 & 0 & 1 & 0 & 2 \end{bmatrix} \cdot \begin{bmatrix} 0 & 0 & 0 & 0 & 0 & 0 & 0 & 1 & 0 \\ 0 & 0 & 0 & 0 & 0 & 0 & 1 & 0 & 0 \\ 0 & 0 & 0 & 0 & 0 & 0 & 0 & 0 & 1 \\ 0 & 1 & 0 & 0 & 0 & 0 & 0 & 0 & 0 \\ 1 & 0 & 0 & 0 & 0 & 0 & 0 & 0 & 0 \\ 0 & 0 & 1 & 0 & 0 & 0 & 0 & 0 & 0 \\ 0 & 0 & 0 & 0 & 1 & 0 & 0 & 0 & 0 \\ 0 & 0 & 0 & 1 & 0 & 0 & 0 & 0 & 0 \\ 0 & 0 & 0 & 0 & 0 & 1 & 0 & 0 & 0 \end{bmatrix} \\ &= \begin{bmatrix} 0 & 0 & 0 & 0 & 0 & 0 & 0 & 0 & 0 \\ 1 & 0 & 2 & 1 & 0 & 2 & 1 & 0 & 2 \\ 2 & 0 & 1 & 2 & 0 & 1 & 2 & 0 & 1 \\ 1 & 1 & 1 & 2 & 2 & 2 & 0 & 0 & 0 \\ 2 & 1 & 0 & 0 & 2 & 1 & 1 & 0 & 2 \\ 0 & 1 & 2 & 1 & 2 & 0 & 2 & 0 & 1 \\ 2 & 2 & 2 & 1 & 1 & 1 & 0 & 0 & 0 \\ 0 & 2 & 1 & 2 & 1 & 0 & 1 & 0 & 2 \\ 1 & 2 & 0 & 0 & 1 & 2 & 2 & 0 & 1 \end{bmatrix}. \end{aligned}$$

Further,

$$Alpha_3(2) = (RM(1) \otimes RM(1)) \cdot (M^{[2]} \cdot P(2)) \cdot (RM(1) \otimes RM(1))^T$$

and

$$Alpha_3(2) = \begin{bmatrix} 0 & 0 & 0 & 0 & 0 & 0 & 0 & 0 & 0 \\ \mathbf{1} & \mathbf{2} & 0 & 0 & 0 & 0 & 0 & 0 & 0 \\ 0 & 0 & 0 & 0 & 0 & 0 & 0 & 0 & 0 \\ \mathbf{1} & 0 & 0 & \mathbf{1} & 0 & 0 & 0 & 0 & 0 \\ 0 & 0 & 0 & 0 & 0 & 0 & 0 & 0 & 0 \\ 0 & 0 & 0 & 0 & 0 & 0 & 0 & 0 & 0 \\ 0 & 0 & 0 & 0 & 0 & 0 & 0 & 0 & 0 \\ 0 & 0 & 0 & 0 & 0 & 0 & 0 & 0 & 0 \\ 0 & 0 & 0 & 0 & 0 & 0 & 0 & 0 & 0 \end{bmatrix},$$

$$Beta_3(2) = \begin{bmatrix} 0 & 0 & 0 & \mathbf{2} & \mathbf{1} & 0 & 0 & 0 & 0 \\ 0 & 0 & 0 & 0 & 0 & 0 & 0 & 0 & 0 \\ 0 & 0 & 0 & 0 & 0 & 0 & 0 & 0 & 0 \\ 0 & 0 & 0 & 0 & 0 & 0 & 0 & 0 & 0 \\ 0 & 0 & 0 & 0 & 0 & 0 & 0 & 0 & 0 \\ 0 & 0 & 0 & 0 & 0 & 0 & 0 & 0 & 0 \\ 0 & 0 & 0 & 0 & 0 & 0 & 0 & 0 & 0 \\ 0 & 0 & 0 & 0 & 0 & 0 & 0 & 0 & 0 \\ 0 & 0 & 0 & 0 & 0 & 0 & 0 & 0 & 0 \end{bmatrix}$$

$Alpha_3(2)$ and $Beta_3(2)$ are clearly disjoint.

3.3 Case $k = 3$, $n = 6$, $p^k = 27$

It becomes evident that in this case (27×27) matrices are the basic components of the system and their explicit representation in the calculations will no longer be reasonable. Calculations were conducted in ©Scilab 6.0.2 [23] and only the results will be presented.

Example 3. *Let*

$$P(3) = \begin{bmatrix} 0 & 0 & 1 \\ 1 & 0 & 0 \\ 0 & 1 & 0 \end{bmatrix} \otimes \begin{bmatrix} 0 & 1 & 0 \\ 1 & 0 & 0 \\ 0 & 0 & 1 \end{bmatrix} \otimes \begin{bmatrix} 0 & 0 & 1 \\ 0 & 1 & 0 \\ 1 & 0 & 0 \end{bmatrix}$$

and

$$G = \begin{bmatrix} 0 & 2 & 1 \end{bmatrix} \otimes \begin{bmatrix} 1 & 0 & 2 \end{bmatrix} \otimes \begin{bmatrix} 2 & 1 & 2 \end{bmatrix} \bmod 3.$$

Then,

$$
\begin{aligned}
Alpha_3(3) \quad & row \quad 2 && = [22000000000 \cdots 00] \\
& row \quad 4 && = [10020000000 \cdots 00] \\
& row \quad 10 && = [10000000010 \cdots 00] \\
Beta_3(3) \quad & row \quad 1 && = [0000000001222110 \cdots 00]
\end{aligned}
$$

All other rows are 0-rows.

Notice that $Alpha_3(3)$ preserved the position of the non-0 rows of $Alpha_3(2)$ and added a new one. $Beta_3(3)$, as expected, is quite different than $Beta_3(2)$ since their only non-0 row transposed equal the RM spectra of the corresponding G^T functions and they were strongly different. Several experiments with other arbitrary permutation matrices produced consistent results in terms of the position of the non-0 rows of $Alpha_3(3)$.

3.4 Case $k = 4$, $n = 8$, $p^k = 81$.

In this case, (81×81) matrices are involved in most calculations and cannot be explicitly disclosed. Only the results of calculations will be reported. Moreover, since it is clear that *Alpha* and *Beta* are disjoint and *Beta* has only one non-0 row in the first position, in what follows only the structure of *Alpha* will be considered. Furthermore, to simplify the representations, permutations and G-functions will be represented as the Kronecker product of elementary components, but consistent result are obtained with arbitrary permutation matrices and arbitrary G-functions.

Example 4. *Let*

$$
P(4) = \begin{bmatrix} 1 & 0 & 0 \\ 0 & 0 & 1 \\ 0 & 1 & 0 \end{bmatrix} \otimes \begin{bmatrix} 0 & 0 & 1 \\ 1 & 0 & 0 \\ 0 & 1 & 0 \end{bmatrix} \otimes \begin{bmatrix} 0 & 1 & 0 \\ 1 & 0 & 0 \\ 0 & 0 & 1 \end{bmatrix} \otimes \begin{bmatrix} 0 & 0 & 1 \\ 0 & 1 & 0 \\ 1 & 0 & 0 \end{bmatrix}
$$

and

$$
G = \begin{bmatrix} 1 & 2 & 2 \end{bmatrix} \otimes \begin{bmatrix} 0 & 2 & 1 \end{bmatrix} \otimes \begin{bmatrix} 1 & 0 & 2 \end{bmatrix} \otimes \begin{bmatrix} 2 & 1 & 2 \end{bmatrix} \bmod 3.
$$

$$
\begin{aligned}
Alpha_3(4) \quad & row \quad 2 && = [22000000000000000000000000000 \cdots 00] \\
& row \quad 4 && = [10020000000000000000000000000 \cdots 00] \\
& row \quad 10 && = [10000000010000000000000000000 \cdots 00] \\
& row \quad 28 && = [00000000000000000000000000020 \cdots 00].
\end{aligned}
$$

All other rows are 0-rows.

k	n	Position of the non-0 rows of $Alpha_3(k)$
1	2	2
2	4	2,4
3	6	2, 4, 10
4	8	2, 4, 10, 28
5	10	2, 4, 10, 28, 82
6	12	2, 4, 10, 28, 82, 244

Table 1: Distribution of the non-0 rows of $Alpha_3(k)$.

Examples with $k = 5$ and $k = 6$ were conducted, but for space reasons cannot be included here. The distribution of the non-0 rows of $Alpha_3(k)$ are summarized in Table 1.

The data in Table 1 support the following (strong) conjectures:

i) There are k non-0 rows in $Alpha_3(k)$,

ii) Let r_j denote the position of some non-0 row of $Alpha$ and r_{j-1} the position of the previous non-0 row. Then for all $2 < j \leq k$ holds: $r_j = 3 \cdot r_{j-1} - 2 = 3(r_{j-1} - 1) + 1$.

4 Reed-Muller spectrum of 5-valued MMF functions

In this section, we discus the structure of the RM spectrum for 5-valued functions.

4.1 Case $k = 1$, $n = 2$, $5^k = 5$.

In the 5-valued domain, the elementary Reed-Muller basis is given by $X_5(1) = \begin{bmatrix} 1 & x & x^2 & x^3 & x^4 \end{bmatrix}$, which in the ring $(Z_5, +, \cdot)$ corresponds to

$$X_5(1) = \begin{bmatrix} 1 & 0 & 0 & 0 & 0 \\ 1 & 1 & 1 & 1 & 1 \\ 1 & 2 & 4 & 3 & 1 \\ 1 & 3 & 4 & 2 & 1 \\ 1 & 4 & 1 & 4 & 1 \end{bmatrix}.$$

As mentioned for (5), its inverse leads to

$$RM_5(1) = \begin{bmatrix} 1 & 0 & 0 & 0 & 0 \\ 0 & 4 & 2 & 3 & 1 \\ 0 & 4 & 1 & 1 & 4 \\ 0 & 4 & 3 & 2 & 1 \\ 4 & 4 & 4 & 4 & 4 \end{bmatrix}, \tag{12}$$

and, as mentioned after (3)

$$M_5 = \begin{bmatrix} 0 & 0 & 0 & 0 & 0 \\ 0 & 1 & 2 & 3 & 4 \\ 0 & 2 & 4 & 1 & 3 \\ 0 & 3 & 1 & 4 & 2 \\ 0 & 4 & 3 & 2 & 1 \end{bmatrix} \tag{13}$$

Recall that

$$\omega F = M^{[k]} \cdot P(k) \oplus (\mathbf{1}^G \otimes G),$$

and with (9),

$$Alpha_5(1) = RM_5(1) \cdot M \cdot P(1) \cdot RM_5(1)^T$$

and

$$Beta_5(1) = RM_5(1) \cdot (\mathbf{1}^G \otimes G) \cdot RM_5(1)^T.$$

Remark 1. *Notice that M_5 has particular properties consistent with M_3: It has a first 0-row. The sum of all row entries and column entries add up to* $0 \bmod 5$. *The first row of $RM_5(1)$ is a* $\begin{matrix} 1 & 0 & \cdots & 0 \end{matrix}$ *row and its bottom row is a constant* 4 *row. Then $RM_5(1) \cdot M_5$ produces a matrix whose first and bottom rows are* 0*-rows. This property, that appears in Alpha, as shown in Example 5 below, is* independent *of the permutation $P(1)$, which mainly affects the entries of the non-*0 *rows.*

In what follows, permutation matrices may be coded with a row vector, where the position of an entry indicates the column of the permutation matrix and the value of that entry indicates the row where the permutation matrix has the entry 1. It is simple to see that coding vectors are permutations of the "identity vector" = $\begin{bmatrix} 1 & 2 & \cdots & 5^k \end{bmatrix}$.

Example 5. *Let*

$$P_5(1) = \begin{bmatrix} 5 & 3 & 4 & 1 & 2 \end{bmatrix}; G = \begin{bmatrix} 4 & 2 & 3 & 0 & 1 \end{bmatrix}.$$

$$\begin{aligned} Alpha_5(1) &= RM_5(1) \cdot M_5 \cdot P_5(1) \cdot RM_5(1)^T \\ &\begin{bmatrix} 1 & 0 & 0 & 0 & 0 \\ 0 & 4 & 2 & 3 & 1 \\ 0 & 4 & 1 & 1 & 4 \\ 0 & 4 & 3 & 2 & 1 \\ 4 & 4 & 4 & 4 & 4 \end{bmatrix} \cdot \begin{bmatrix} 0 & 0 & 0 & 0 & 0 \\ 0 & 1 & 2 & 3 & 4 \\ 0 & 2 & 4 & 1 & 3 \\ 0 & 3 & 1 & 4 & 2 \\ 0 & 4 & 3 & 2 & 1 \end{bmatrix} \\ &\quad \cdot \begin{bmatrix} 0 & 0 & 0 & 1 & 0 \\ 0 & 0 & 0 & 0 & 1 \\ 0 & 1 & 0 & 0 & 0 \\ 0 & 0 & 1 & 0 & 0 \\ 1 & 0 & 0 & 0 & 0 \end{bmatrix} \cdot \begin{bmatrix} 1 & 0 & 0 & 0 & 0 \\ 0 & 4 & 2 & 3 & 1 \\ 0 & 4 & 1 & 1 & 4 \\ 0 & 4 & 3 & 2 & 1 \\ 4 & 4 & 4 & 4 & 4 \end{bmatrix}^T \\ &= \begin{bmatrix} 0 & 0 & 0 & 0 & 0 \\ 4 & 0 & 0 & 3 & 0 \\ 0 & 0 & 0 & 0 & 0 \\ 0 & 0 & 0 & 0 & 0 \\ 0 & 0 & 0 & 0 & 0 \end{bmatrix}. \end{aligned}$$

$$\begin{aligned} Beta_5(1) row\ 1 &= (RM_5(1) \cdot G^T)^T \\ &= \left(\begin{bmatrix} 1 & 0 & 0 & 0 & 0 \\ 0 & 4 & 2 & 3 & 1 \\ 0 & 4 & 1 & 1 & 4 \\ 0 & 4 & 3 & 2 & 1 \\ 4 & 4 & 4 & 4 & 4 \end{bmatrix} \cdot \begin{bmatrix} 4 \\ 2 \\ 3 \\ 0 \\ 1 \end{bmatrix} \right)^T = \begin{bmatrix} 4 & 0 & 0 & 3 & 0 \end{bmatrix}. \end{aligned}$$

Examples with other permutations and G functions gave consistent results.

4.2 Case $k = 2$, $n = 4$, $5^k = 25$.

For these values of k and n,

$$\begin{aligned} RM_5(2) &= RM_5(1) \otimes RM_5(1), \\ M_5^{[2]} &= (M_5 \otimes \mathbf{1}^M) \oplus (\mathbf{1}^M \otimes M_5). \end{aligned}$$

Example 6. *Let* $P(2) = P_1 \otimes P_2$*, with*

$$\begin{aligned} P_1 &= \begin{bmatrix} 5 & 3 & 4 & 1 & 2 \end{bmatrix}, \\ P_2 &= \begin{bmatrix} 3 & 5 & 2 & 4 & 1 \end{bmatrix}. \end{aligned}$$

The Kronecker product is meant to be of the corresponding permutation matrices; not of their coding vectors.

$$G = \begin{bmatrix} 1 & 0 & 3 & 2 & 4 \end{bmatrix} \otimes \begin{bmatrix} 3 & 2 & 1 & 2 & 3 \end{bmatrix} \bmod 5.$$

Notice that $P(2)$ *is a* (25×25) *matrix and* G *has a length of* 25.

$$Alpha_5(2) = RM_5(2) \cdot M_5^{[2]} \cdot P(2) \cdot RM_5(2)^T.$$

$$\begin{aligned} Alpha_5(2) \quad row \quad 2 &= [220000 \cdots 00] \\ row \quad 6 &= [400000 \cdots 00] \end{aligned}$$

All other rows are 0*-rows.*

Examples with other arbitrary permutations and G functions gave consistent results.

4.3 Case $k = 3$, $n = 6$, $5^k = 125$.

For these values of k and n, it is

$$\begin{aligned} RM_5(3) &= RM_5(1) \otimes RM_5(2), \\ M_5^{[3]} &= (M_5^{[2]} \otimes \mathbf{1}^M) \oplus (\mathbf{1}^{M^{[2]}} \otimes M_5). \end{aligned}$$

Example 7. *Let* P_1 *and* P_2 *be as in Example 6 and chose* $P_3 = \begin{bmatrix} 5 & 1 & 4 & 2 & 3 \end{bmatrix}$. *Then, let* $P_5(3) = P_1 \otimes P_2 \otimes P_3$. *Notice that* $P_5(3)$ *is a* (125×125) *matrix.*

$$Alpha_5(3) = RM_5(3) \cdot M_5^{[3]} \cdot P_5(3) \cdot RM_5(3)^T.$$

$$\begin{aligned} Alpha_5(3) \quad row \quad 2 &= [412300 \cdots 000 \cdots 00] \\ row \quad 6 &= [200002 \cdots 000 \cdots 00] \\ row \quad 26 &= [400000 \cdots 030 \cdots 00]. \end{aligned}$$

All other rows are 0*-rows.*

Additional examples with other arbitrary permutations gave consistent results.

k	n	Position of the non-0 rows of $Apha_5(k)$
1	2	2
2	4	2, 6
3	6	2, 6, 26
4	8	2, 6, 26, 126

Table 2: The non-0 rows of $Alpha_5(k)$.

4.4 Case $k = 4$, $n = 8$, $5^k = 625$.

For these values of k and n, it is

$$\begin{aligned} RM_5(4) &= RM_5(2) \otimes RM_5(2), \\ M_5^{[4]} &= (M_5^{[2]} \otimes \mathbf{1}^{M[2]}) \oplus (\mathbf{1}^{M[2]} \otimes M_5^{[2]}). \end{aligned}$$

Example 8. *Let P_1, P_2, and P_3 be as in Example 7 and chose*

$$P_4 = \begin{bmatrix} 4 & 2 & 1 & 3 & 5 \end{bmatrix}.$$

Then, let $P_5(4) = P_1 \otimes P_2 \otimes P_3 \otimes P_4$. Notice that $P_5(4)$ is a (625×625) matrix.

$$Alpha_5(4) = RM_5(4) \cdot M_5^{[4]} \cdot P_5(4) \cdot RM_5(4)^T.$$

Since the Alpha matrix is sparse, the following notation will be used for its non-0 rows: v_w will denote a non-0 entry, where v indicates the value of an entry and w tells "where", i.e., the position of that non-0 entry.

$$\begin{array}{llrl} Alpha_5(4) & row & 2 & = [3_1, 4_2, 2_3, 2_4] \\ & row & 6 & = [4_1, 1_6, 2_{11}, 3_{16}] \\ & row & 26 & = [2_1, 2_{26}] \\ & row & 126 & = [4_1, 3_{376}]. \end{array}$$

Notice that to try to analyze the case $k = 5$, it would require calculations with (3125×3125) matrices, but this is beyond the possibilities of the available computing environment. A different representation would be needed. This is however outside the scope of this paper.

The available *Alpha* data is summarized in Table 2.

The available data support the following conjectures, closely related to those stated for $Alpha_3(k)$:

i) There are k non-0 rows in $Alpha_5(k)$,

ii) Let r_j denote the position of some non-0 row of $Alpha$ and r_{j-1} the position of the previous non-0 row. Then, for all $2 \leq j \leq k$ holds: $r_j = 5 \cdot r_{j-1} - 4 = 5(r_{j-1} - 1) + 1$.

5 Reed-Muller spectrum of 7-valued MMF functions

Notice that $7^2 = 49$, $7^3 = 343$ and $7^4 = 2,401$. Therefore, only the cases $k < 4$ can be reported.

The Reed-Muller basis is given by

$$X_7(1) = \left[\begin{array}{ccccccc} 1 & x & x^2 & x^3 & x^4 & x^5 & x^6 \end{array} \right],$$

which in $(Z_7, +, \cdot)$ becomes

$$X_7(1) = \left[\begin{array}{ccccccc} 1 & 0 & 0 & 0 & 0 & 0 & 0 \\ 1 & 1 & 1 & 1 & 1 & 1 & 1 \\ 1 & 2 & 4 & 1 & 2 & 4 & 1 \\ 1 & 3 & 2 & 6 & 4 & 5 & 1 \\ 1 & 4 & 2 & 1 & 4 & 2 & 1 \\ 1 & 5 & 4 & 6 & 2 & 3 & 1 \\ 1 & 6 & 1 & 6 & 1 & 6 & 1 \end{array} \right].$$

Therefore,

$$RM_7(1) = \left[\begin{array}{ccccccc} 1 & 0 & 0 & 0 & 0 & 0 & 0 \\ 0 & 6 & 3 & 2 & 5 & 4 & 1 \\ 0 & 6 & 5 & 3 & 3 & 5 & 6 \\ 0 & 6 & 6 & 1 & 6 & 1 & 1 \\ 0 & 6 & 3 & 5 & 5 & 3 & 6 \\ 0 & 6 & 5 & 4 & 3 & 2 & 1 \\ 6 & 6 & 6 & 6 & 6 & 6 & 6 \end{array} \right], \qquad (14)$$

moreover,

$$M_7 = \left[\begin{array}{ccccccc} 0 & 0 & 0 & 0 & 0 & 0 & 0 \\ 0 & 1 & 2 & 3 & 4 & 5 & 6 \\ 0 & 2 & 4 & 6 & 1 & 3 & 5 \\ 0 & 3 & 6 & 2 & 5 & 1 & 4 \\ 0 & 4 & 1 & 5 & 2 & 6 & 3 \\ 0 & 5 & 3 & 1 & 6 & 4 & 2 \\ 0 & 6 & 5 & 4 & 3 & 2 & 1 \end{array} \right]. \qquad (15)$$

5.1 Case $k = 1$, $n = 2$, $7^k = 7$

The following example illustrates this selection of k and n for $p = 7$.

Example 9. *Let* $P_7(1) = \left[\begin{array}{ccccccc} 2 & 4 & 6 & 5 & 1 & 2 & 7 \end{array}\right]$. *Then,*

$$Alpha_7(1) = RM_7(1) \cdot M_7 \cdot P_7(1) \cdot RM_7(1)^T.$$

$$Alpha_7(1) \quad row \quad 2 = \left[\begin{array}{ccccccc} 2 & 2 & 5 & 3 & 1 & 4 & 0 \end{array}\right].$$

All other rows are 0-rows.

5.2 Case $k = 2$, $n = 4$, $7^k = 49$.

In this case,

$$\begin{aligned} RM_7(2) &= RM_7(1) \otimes RM_7(1), \\ M_7^{[2]} &= (M_7 \otimes \mathbf{1}^M) \oplus (\mathbf{1}^M \otimes M_7). \end{aligned}$$

Example 10. *Let*

$$\begin{aligned} P_7(2) &= P_7(1) \otimes P_7(1), \\ Alpha_7(2) &= RM_7(2) \cdot M_7^{[2]} \cdot P_7(2) \cdot RM_7(2)^T. \end{aligned}$$

$$\begin{aligned} Alpha_7(2) \quad row \quad 2 &= [2253140 \cdots 00] \\ row \quad 8 &= [2_1, 2_8, 5_{15}, 3_{22}, 1_{29}, 4_{36}]. \end{aligned}$$

All other rows are 0-rows.

5.3 Case $k = 3$, $n = 6$, $7^k = 343$.

In this case,

$$\begin{aligned} RM_7(3) &= RM_7(1) \otimes RM_7(2), \\ M_7^{[3]} &= (M_7 \otimes 1^{M^{[2]}}) \oplus (\mathbf{1}^M \otimes M_7^{[2]}). \end{aligned}$$

Example 11. *Let*

$$\begin{aligned} P_7(3) &= P_7(1) \otimes P_7(2), \\ Alpha_7(3) &= RM_7(3) \cdot M_7^{[3]} \cdot P_7(3) \cdot RM_7(3)^T. \end{aligned}$$

k	n	Position of the non-0 rows of $Alpha_7(k)$
1	2	2
2	4	2, 8
3	6	2, 8, 50

Table 3: The non-0 rows of $Alpha_7(k)$.

$$\begin{aligned} Alpha_7(3) \quad & row \quad 2 &&= [2253140\cdots 00] \\ & row \quad 8 &&= [2_1, 2_8, 5_{15}, 3_{22}, 1_{29}, 4_{36}]. \\ & row \quad 50 &&= [2_1, 2_{50}, 5_{99}, 3_{148}, 1_{197}, 4_{246}]. \end{aligned}$$

The distribution of the non-0 rows of $Alpha_7(k)$ are shown in Table 3.

This data, albeit a small amount, supports conjectures closely related to those for the cases with $p = 3$ and $p = 5$.

i) There are k non-0 rows in $Alpha_7(k)$,

ii) Let r_j denote the position of some non-0 row of $Alpha$ and r_{j-1} the position of the previous non-0 row. Then, for all $2 \leq j \leq k$ holds: $r_j = 7 \cdot r_{j-1} - 6 = 7(r_{j-1} - 1) + 1$.

The first conjecture is general, since it is supported by the three considered values of p.

The second conjecture may be generalized as:
For all $2 \leq j \leq k$ holds

$$r_j = p \cdot r_{j-1} - (p-1) = p(r_{j-1} - 1) + 1. \tag{16}$$

6 Induction proof for the conjectures.

Preliminaries

Direct calculations show that $RM_p(1) \cdot \mathbf{1}^M \cdot RM_p(1)^T = diag(100000\ldots 0) = \mathbf{1}^m$. Furthermore, it is obvious that for any permutation and any k, $\mathbf{1}^{M[k]} \cdot P_p(k) = \mathbf{1}^{M[k]}$.

Induction basis

Assume that for some $u > 2$ there are u non-0 rows in $Alpha_p(u)$ and their distribution follows the conjectures.

Induction step

$$Alpha_p(u+1) = RM_p(u+1) \cdot M_p^{[u+1]} \cdot P_p(u+1) \cdot RMp(u+1)^T.$$

Recall that $M_p^{[u+1]} = (M_p^{[u]} \otimes \mathbf{1}^M) \oplus (\mathbf{1}^{M[u]} \otimes M_p)$ and assume that (to simplify the presentation of the proof) $P_p(u+1) = P_p(u) \otimes P_p(1)$ for some $P_p(1)$. Then,

$$\begin{aligned}
M_p^{[u+1]} \cdot P_p(u+1) &= \langle (M_p^{[u]} \otimes \mathbf{1}^M) \oplus (\mathbf{1}^{M^{[u]}} \otimes M_p) \rangle \langle P_p(u) \otimes P_p(1) \rangle \\
&= (M_p^{[u]} \otimes \mathbf{1}^M) \cdot (P_p(u) \otimes P_p(1)) \\
&\quad \oplus (\mathbf{1}^{M^{[u]}} \otimes M_p) \cdot (P_p(u) \otimes P_p(1)) \\
&= M_p^{[u]} \cdot P_p(u) \otimes \mathbf{1}^M \cdot P_p(1) \oplus \mathbf{1}^{M^{[u]}} \cdot P_p(u) \otimes M_p \cdot P_p(1) \\
&= M_p^{[u]} \cdot P_p(u) \otimes \mathbf{1}^M \oplus \mathbf{1}^{M^{[u]}} \otimes M_p \cdot P_p(1).
\end{aligned}$$

Hence,

$$\begin{aligned}
Alpha_p(u+1) &= (RM_p(u) \otimes RM_p(1)) \cdot \langle M_p^{[u]} \cdot P_p(u) \otimes \mathbf{1}^M \\
&\quad \oplus \mathbf{1}^{M^{[u]}} \otimes M_p \cdot P_p(1) \rangle \cdot (RM_p(u) \otimes RM_p(1))^T \\
&= (RM_p(u) \cdot M_p^{[u]} \cdot P_p(u) \cdot RM_p(u)^T) \otimes (RM_p(1) \cdot \mathbf{1}^M \cdot RM_p(1)^T) \\
&\quad \oplus (RM_p(u) \cdot \mathbf{1}^{M^{[u]}} \cdot RM_p(u)^T) \\
&\quad \otimes (RM_p(1) \cdot M_p \cdot P_p(1) \cdot RM_p(1)^T) \qquad (17) \\
&= Alpha_p(u) \otimes \mathbf{1}^m \oplus (\mathbf{1}^m)^{\otimes u} \otimes Alpha_p(1).
\end{aligned}$$

Recall that $\mathbf{1}^m$ is a $(p \times p)$ matrix with a single non-0 entry at the position $(1, 1)$. $Alpha_p(u)$ is a $(p^u \times p^u)$ matrix, where the position of its u (non-0) rows follows the corresponding conjecture. Then, $Alpha_p(u) \otimes \mathbf{1}^m$ is a $(p^{u+1} \times p^{u+1})$ matrix such that every entry of $Alpha_p(u)$ is replaced by a $\mathbf{1}^m$ matrix scaled by the value of the corresponding entry. In other words, $Alpha_p(u) \otimes \mathbf{1}^m$ is a matrix whose rows are ordered in blocks of p rows. The first row of each block depends on the row of $Alpha_p(u)$ associated to that block. The remaining rows of the block will be 0-rows. Notice that therefore, the first block of $Alpha_p(u) \otimes \mathbf{1}^m$ will be a 0-block. However, $(\mathbf{1}^m)^{\otimes u}$ is a $(p^u \times p^u)$ matrix such that its only non-0 entry is at the position $(1, 1)$. Therefore, $(\mathbf{1}^m)^{\otimes u} \otimes Alpha_p(1)$ copies $Alpha_p(1)$ at its left upper corner. With (17) this provides a non-0 row at the 2nd position for $Alpha_p(u+1)$, as needed.

Let the positions of the rows of $Alpha_p(u)$ be first assigned to the blocks of $Alpha_p(u+1)$ and recall that each row of $Alpha_p(u)$ became the first row of a block of $Alpha_p(u+1)$. Notice that with this construction, $Alpha_p(u+1)$ will have (only) the u non-0 rows of $Alpha_p(u)$. Recall, however, that as detailed above, the second term of (17) provides the "missing" non-0 row at the second position and then $Alpha_p(u+1)$ comprises $u+1$ non-zero rows (which proves the first conjecture.) Now, if the first row of the r_j-th block of $Alpha_p(u+1)$ is a non-0 row, it has r_{j-1} preceding blocks, i.e., $p(r_j-1)$ preceding rows. Therefore, its own row-position is $p(r_j-1)+1 = pr_j-(p-1)$. This proves the second generalized conjecture. □

7 Closing Remarks

We have considered p as an odd prime and have proven that the Reed-Muller spectra of p-valued Maiorana-McFarland (MMF) bent functions have special properties, which remain valid for different values of p and $n = 2k$. In [14] we had already proven that an analog to (16), (with $p = 2$) holds for Maiorana-McFarland bent functions in the Boolean domain. In both cases, experiments done with bent functions which do not belong to the respective Maiorana-McFarland classes lead to Reed-Muller spectra with a different number and distribution of non-0 rows. This supports the conclusion, that the disclosed results provide a characterization for MMF bent functions. Without this, to find out whether a p-valued bent function is MMF or not, besides checking for necessary conditions [12], reversing Eq. (3) would be needed. This has however, a higher complexity than the product of the Reed-Muller transform matrix and the value-matrix of the function.

References

[1] Chrestenson H. E., "A class of generalized Walsh functions," *Pacific J. of Math.*, **5**, (5), 17-31, 1955.

[2] Green D. H., Taylor I. S., "Multiple-valued switching circuit design by means of generalized Reed-Muller expansions," *Digital Processes*, **2**, 63-81, 1976.

[3] Horn R. A., Johnson Ch. R., *Topics in Matrix Analysis.* Cambridge University Press, New York, 1991.

[4] Kumar P. V., Scholtz R. A., Welch L. R., "Generalized bent functions and their properties," *J. Combinatorial Theory*, Vol. A, No. 40, 90-107, 1985.

[5] Luis M., Moraga C., "Functions with flat Chrestenson spectra," *Proc.19th Int. Symp. Multiple-valued Logic*, Guangzhou, China, 406-413, IEEE Press, 1989.

[6] Maiorana J., "A Class of Bent Functions," R41 Technical Paper, August 1970

[7] McFarland R., “A discrete Fourier theory for binary functions,” R41 Technical Paper; June 1971.

[8] Mesnager S., *Bent Functions. Fundamentals and Results.*, Springer, 2016.

[9] Moraga C., Stanković R. S., Stanković M., Stojković S., “Contribution to the study of ternary bent functions,” in *Proc. 43rd Int. Symp. on Multiple-Valued Logic*, 340-345, IEEE Press, 2013.

[10] Moraga C., Stanković M., Stanković R. S., Stojković S., “The Maiorana Method to generate ternary bent functions revisited,” in *Proc. 44th Int. Symp. on Multiple-Valued Logic*, 19-24, IEEE Press, 2014.

[11] Moraga C., Stanković M., Stanković R. S., “Multiple-valued functions with bent Reed-Muller spectra,” (B. Steinbach, Ed.) Problems and new Solutions in the Boolean Domain, 309-324, Cambridge Scholar Publishing, Newcastle upon Tyne, UK, 2016.

[12] Moraga C., Stanković R. S., Stanković M., “New properties of the Maiorana-McFarland Ternary Bent Functions,” In *Proc. 52nd. Int. Symp. on Multiple-Valued Logic*, 56-61, IEEE Xplore, 2022.

[13] Moraga C., Stanković R. S., Stanković M., “Reed-Muller-Fourier Spectra of p-valued MMF Bent Functions,” In *Proc. 53rd Int. Symposium on Multiple-Valued Logic*, 64-69, IEEE Press, 2023.

[14] Moraga C., Stanković R. S., Stanković M., “Properties of the Reed-Muller Spectrum of Maiorana-McFarland Boolean Bent Functions,” In *Proc. Int. Workshop on Boolean Functions*, 47-61, Press University of Bremen, 2022.

[15] Muller D.E., “Application of Boolean algebra to switching circuit design and error detection,” *IRE Trans. Electron. Comp.*, 1, 6–12, 1954.

[16] Reed I. S., “A class of multiple-error-correcting codes and their decoding scheme,” *IRE Trans. Information Theory*, 3, 6–12, 1954.

[17] Rothaus O.S., “On ‘bent’ Functions,” *J. of Combinatorial Theory*, Series A, 20, 300-305, 1976.

[18] Stanković R. S., Moraga C., Astola J., “Reed-Muller expressions in the last decade,” *Proc. International Workshop on Theory and Applications of Reed Muller Expressions*, 7-26, University of Mississippi, USA, 2001.

[19] Stanković M. M., Moraga C., Stanković R. S., “Construction of ternary bent functions by spectral invariant operations in the Generalized Reed-Muller domain,” In *Proc. 48th Int. Symposium on Multiple-Valued Logic*, 235-240, IEEE Press, 2018.

[20] Tokareva N., *Bent Functions. Results and Applications to Cryptography.* Elsevier – Academic Press, Amsterdam, 2015.

[21] Vilenkin N. Ya., Agaev G. N., Dzafarli G. M., “Towards a theory of multiplicative orthogonal systems of functions,” *DAN Azerb. SSR*, 18, (9), 3-7, 1962.

[22] Walsh, J. L., “A closed set of orthogonal functions,” *Amer. J. Math.*, 55, 5-24, 1923.

[23] www.scilab.org

[24] Zhegalkin, I. L., “O tekhnyke vychyslenyi predlozhenyi v symbolytscheskoi logykye,” *Math. Sb.*, Vol. 34, 9-28, 1927. In Russian.

[25] Zhegalkin, I. L., "Aritmetizatiya symbolytscheskoi logyky," *Math. Sb.*, Vol. 35, 311-377, 1928. In Russian.

www.ingramcontent.com/pod-product-compliance
Lightning Source LLC
Chambersburg PA
CBHW051214200326
41519CB00025B/7105
9781848904446